H.H. Rossi M. Zaider
Microdosimetry and Its Applications

Springer

Berlin
Heidelberg
New York
Barcelona
Budapest
Hong Kong
London
Milan
Paris
Santa Clara
Singapore
Tokyo

H.H. Rossi   M. Zaider

Microdosimetry and Its Applications

With 128 Figures

 Springer

H.H. Rossi
Professor emeritus
Columbia University
105 Larchdale Avenue
Upper Nyack, NY 10960
USA

Dr. M. Zaider
Columbia University
College of Physicians and Surgeons
Radiation Oncology
622 West 168th Street
New York, NY 10032
USA

Cataloging-in-Publication Data applied for

Die Deutsche Bibliothek - CIP-Einheitsaufnahme

Rossi, H. H.:
Microdosimetry and its applications / H. H. Rossi ; M. Zaider.
- Berlin ; Heidelberg ; New York ; Barcelona ; Budapest ;
Hong Kong ; London ; Milan ; Paris ; Santa Clara ; Singapore ;
Tokyo : Springer, 1996
  ISBN 3-540-58541-9
NE: Zaider, M.:

ISBN 3-540-58541-9 Springer-Verlag Berlin Heidelberg New York

© Springer-Verlag Berlin Heidelberg 1996
Printed in Germany

Production: PRODUserv Springer Produktions-Gesellschaft, Berlin
Dataconversion: D. Schüler, Berlin
Cover-layout: MetaDesign plus GmbH, Berlin

SPIN 10066854   51/3020 5 4 3 2 1 0   Printed on acid-free paper

Preface

Microdosimetry originated more than 35 years ago when the senior author studied energy deposition in small irradiated masses and formulated what is now termed Regional Microdosimetry. A.M. Kellerer developed the further concepts of Structural Microdosimetry. Microdosimetry and its applications have been the subject of an extensive literature. This includes several hundred papers which have appeared in the Proceedings of what to date have been eleven Symposia on Microdosimetry. General reviews are contained in chapters of books and in a journal dealing with a broader range of subjects. The International Commission on Radiation Units and Measurements has produced Report 36 on Microdosimetry. The form of these publications limited their scope and in this work it is our aim to provide a more comprehensive account.

Dealing extensively with interdisciplinary matters (from atomic and solid state physics to integral geometry and molecular biology) we were confronted with the standard problem of presenting material in a manner that makes it comprehensible to readers with diverse backgrounds, without providing a superficial treatise. Although some readers may find it too difficult to absorb the entire contents of some chapters, they should be able to gain substantial information from introductory sections and from data presented. Occasional repetitions, including identical formulae, have reduced the need for cross-references between chapters.

With the exception of chapter III it is recommended to read the book sequentially; this appeared to us to be the logical way of learning microdosimetry. The conceptual framework of microdosimetry is introduced in the first two chapters. At this stage of the presentation the details of energy deposition in matter are unimportant. The fourth and fifth chapter are the twin workhorses of microdosimetry. The fourth chapter is about the art of measuring microdosimetric spectra. It gives details on the construction and operation of microdosimetric detectors. It also makes aware both the experimentalist and the theorist that what one measures may sometimes be different from the actual pattern of energy deposition. The material in Chapter V, the theoretical companion of the preceding chapter, comes as a result of the tremendous progress made during the past 20 years or so in obtaining cross sections for the interaction of charged particles with matter, and of the subsequent effort made to integrate these into sophisticated Monte Carlo transport codes that simulate the passage of particles through structured or unstructured matter. This chapter also brings forth the two complementary descriptions of microdosimetric events: *regional* microdosimetry and *structural* microdosimetry. The applications of microdosimetry are described in the last two chapters. The selection of topics here reflects the research interests of the authors and also the need to provide this information within the limitations of a single volume. As a result the list of topics here should be seen as illustrative rather than comprehensive.

Chapter III is largely self contained (the only exception is section III.6.2 that makes use of material treated in II.2). The general intention in this chapter was to collect known mathematical expressions that describe the interaction of radiation with matter. An unusual and perhaps to some surprising feature is the Appendix to this chapter which contains a rather advanced formal treatment of the scattering formalism. The reasons for adding this Appendix are twofold: Firstly, the material presented here is important for understanding (and perhaps being able to derive) many of the equations used in Chapters III and V. More importantly however, the treatment presented here is *phase-independent* (in fact it originates in many-body physics) and as such it is relevant to the gradual replacement of the current gas-based microdosimetry with the study of (relevant) energy depositions in realistic targets such as DNA and protein chains of known sequence, or the structured water that surrounds these biomolecules when in cellular environment. Some examples of these newer results are given in the book.

We have adopted various common simplifications of terminology. Thus the proper term "absorbed dose" is frequently abbreviated to "dose", especially in such expressions as "dose rate" or "dose effect curve". We have found it necessary to employ the same symbol for different quantities if only because of general practice. Thus Q has traditionally stood for both the change of rest mass in nuclear reactions and the quality factor in radiation protection. An effort was made to avoid the converse situation where different symbols refer to the same quantity.

Several chapters from this book have been used by one of us (MZ) to teach microdosimetry to engineering students enrolled in a Master's level graduate program in Medical Physics and Health Physics at Columbia University. In particular, material from Chapter IV is used in conjunction with a laboratory course where students are being familiarized, among other things, with experimental microdosimetry. In the following we recommend a syllabus for a one semester (13-week) course based on material from this book.

Lecture I: Chapters I and II
Lecture II: Sections III.1 to III.3
Lecture III: Section III.4
Lecture IV: Sections III.5 and III.6
Lecture V: Sections IV.1 to IV.3
Lecture VI: Sections IV.4 to IV.7
Lecture VII: Section V.2

Lecture VIII: Section V.3
Lecture IX: Section V.4
Lecture X: Sections VI.1.1 and VI.1.2
Lecture XI: Section VI.1.3
Lecture XII: Section VI.1.4
Lecture XIII: Section VI.2 and VI.3

New York, 1994   Harald H. Rossi and Marco Zaider

Acknowledgments

We are grateful for the help of several fellow scientists. In particular, we wish to acknowledge the importance of many discussions with Prof. A.M. Kellerer who has made numerous and profound contributions to microdosimetry and its applications in radiobiology and radiation protection.

We are much indebted to Dr. D. Srdoc and Prof. A. Wambersie whose expert comments resulted in improvements of the chapters dealing with experimental microdosimetry and microdosimetry in radiotherapy. Help in the preparation of individual sections was also given by Prof. J.F. Dicello, Dr. P. Goldhagen, Dr. G. Luxton and Prof. P.J. McNulty. While the design of the proportional counters described here originated with the senior author they were frequently improved and skillfully executed by Mr. Rudolph Gand - later in collaboration with his worthy successor Gary W. Johnson who also provided the photographs of these instruments.

The final text of the book was cross checked in its entirety for possible inconsistencies of notations and/or textual errors by Ian H. Zaider. He is also responsible for patiently and efficiently renumbering all the equations and figures when single chapters were consolidated in one draft. The efficient help of Ms. Emilia Schneider was essential.

Our deep thanks and appreciation go to all those mentioned above.

New York, 1994

Table of Contents

I   Introduction 1

I.1   The Role of Microdosimetry 1
I.2   The Transfer of Energy from Ionizing Radiation to Matter 2
I.3   Stochastic Quantities 6
I.4   Spatial Aspects of Microdosimetry.. 10
I.5   Temporal Aspects of Microdosimetry 13

II   Microdosimetric Quantities and their Moments 17

II.1   Definitions 17
II.2   Microdosimetric Distributions and their Moments...................... 18
II.3   Representations of Microdosimetric Distributions 23
II.4   Experimental versus Calculated Microdosimetric Distributions 26

III Interactions of Particles with Matter 28

III.1   Overview 28
III.2   Quantities and Terms Relating to the Interaction Between
         Projectiles and Targets.. 29
III.3   Kinematics of the Scattering Process... 34
III.4   Sources of Charged Particles 39
         III.4.1   Photon-interaction Cross Sections............................... 39
         III.4.2   Neutron-interaction Cross Sections 44
         III.4.3   Charged Particles as Sources of other Charged Particles 46
III.5   Microscopic Description of the Electromagnetic Interaction of
         Charged Particles with Matter 48
         III.5.1   Theoretical Outline.. 49
         III.5.2   Experimental Data on the Energy Loss Function 50
III.6   The Interaction of Charged Particles with Bulk Matter.............. 52
         III.6.1   The Stopping Power of the Medium............................ 52
         III.6.2   Statistical Fluctuations of the Energy Lost by Charged
                    Particles 56
         III.6.3   Range and Range Straggling 57

III.7   Appendix: Formal Treatment of the Interaction of Charged
        Particles with Matter..59
        III.7.1   Scattering Formalism...59
        III.7.2   The Dielectric Response Function...66
        III.7.3   Theoretical Calculations of the Energy Loss Function...............70
                  III.7.3.1   Drude-function Expansions of (q,ω).............................70
                  III.7.3.2   Random-phase Approximation (RPA) for (q,ω).........71
                  III.7.3.3   Ab initio Calculations of (q,ω).......................................72

IV   Experimental Microdosimetry..73

IV.1   The Site Concept 73
IV.2   Fluctuations in Regional Microdosimetry...78
IV.3   Measurements in Regional Microdosimetry..83
        IV.3.1   General Considerations..83
        IV.3.2   The Proportional Counter...85
        IV.3.3   Energy loss versus Ionization..86
        IV.3.4   Gas Multiplication...89
        IV.3.5   The Wall Effect 96
        IV.3.6   Tissue Equivalent Materials 100
        IV.3.7   Counter Designs 103
        IV.3.8   Gas Supply..110
        IV.3.9   Electronics 111
        IV.3.10   Calibration 114
        IV.3.11   Resolution...115
IV.4   Measured Distributions of Lineal Energy...117
        IV.4.1   General Comments 117
        IV.4.2   Neutrons 119
        IV.4.3   Photons 123
        IV.4.4   Electrons..126
        IV.4.5   Ions 129
        IV.4.6   Pions 133
IV.5   Measurement of Distributions of Specific Energy.................................134
        IV.5.1   General Comments 134
        IV.5.2   The Variance Method...135
IV.6   Measurement of LET Distributions 139
IV.7   Appendix: The V Effect 142

V   Theoretical Microdosimetry 148

V.1   A Diversion in Geometric Probability..148
V.2   Monte Carlo Simulation of Charged-Particle Tracks 152

V.2.1   A Brief Visit to Monte Carlo Sampling 152
V.2.2   Geometrical Randomness 157
V.2.3   An Illustration: Monte Carlo Simulation of Electron Tracks 158
V.3   Calculation of Microdosimetric Spectra 166
V.3.1   Analytic Methods 166
V.3.2   Monte Carlo Methods 171
V.3.3   Microdosimetric Spectra for Combined Radiations 173
V.4   Methods for Obtaining Proximity Functions 176
V.4.1   Proximity Functions for Simple Geometric Objects 176
V.4.2   Proximity Functions for Amorphous Tracks. 179
V.4.3   Proximity Functions from Experimental Data 184
     V.4.3.1 $t(x)$ and y_D 185
     V.4.3.2   Proximity Functions for Diffused Charged
                     Particle Tracks 188
     V.4.3.3   Proximity Functions obtained from
                     Cloud-Chamber Data 190
V.5   The Informational Content of the Moments of the
     Microdosimetric Distributions 198
V.6   Appendix: The Maximum Entropy Principle 201

VI   Applications of Microdosimetry in Biology 205

VI.1   Radiobiology 205
VI.1.1   Introduction 205
VI.1.2   Microdosimetric Constraints on Biophysical Models 206
VI.1.3   Empirical Data in Radiation Biology 216
VI.1.4   The Theory of Dual Radiation Action 229
VI.1.5   Other Topics in Dual Radiation Action Theory 241
VI.1.6   DNA-lesion Theory of Radiation Action 247
VI.2   Radiotherapy 250
VI.2.1   General Considerations 250
VI.2.2   Microdosimetric Distributions 253
VI.3   Radiation Protection 264
VI.3.1   Quantities 264
VI.3.2   General Considerations Regarding Measurements 269
VI.3.3   Measurements of the "Counter" Dose Equivalent 272
VI.3.4   Measurement of Operational Quantities 274
VI.3.5   Specific Quality Functions 275

VII   Other Applications 279

VII.1   Microdosimetry and Radiation Chemistry 279
VII.2   Radiation Effects on Microelectronics 291

VII.2.1 Appendix: Example 297
VII.3 Microdosimetry and Thermoluminescence 298

References 301

Subject Index 317

Chapter I
Introduction

I.1 The Role of Microdosimetry

In its action on matter, including living matter, ionizing radiation is uniquely efficient because it transfers energy to atoms in a highly concentrated form. The average energy absorbed per unit mass of irradiated medium, the absorbed dose, is minute compared to the energy densities that elicit comparable effects by other physical agents. The effectiveness of ionizations is further enhanced by their association in the tracks of charged particles. Depending on its *microscopic* distribution the energy required for a given level of biological injury may vary as much as a hundredfold between different ionizing radiations. Thus while the absorbed dose is a useful and standard quantity in the specification of irradiation, effects depend on the *pattern* in which a given amount of energy is deposited in the irradiated medium. A knowledge of such energy distributions is required not only in any explanations of the relative effectiveness of different kinds of ionizing radiation but it also can be expected to provide insight into the action of ionizing radiation in general. This has been recognized since the earliest days of radiobiology and led to the fundamental contributions by such investigators as Crowther, Dessauer, Timofeff-Resowsky, Zirkle and Lea.

Microdosimetry, which was developed as a system of concepts as well as of physical quantities and their measurement, is the systematic study and quantification of the *spatial* and *temporal* distribution of absorbed energy in irradiated matter. Although it can, and has been, utilized in the interpretation of a variety of radiation effects, its principal utility has been in the field of radiobiology. This aspect will be emphasized in this book and is addressed in much of this introduction.

The absorbed energy is transformed into initial molecular changes or degraded into heat in a period that is generally less than 1 microsecond. In this brief time interval many highly complex interactions occur before reasonably stable molecular species are produced but the effects under consideration may not develop for periods that can be much longer; in the case of carcinogenesis they may be decades. Despite these intricacies, microdosimetry can provide clues to the understanding of early processes that determine the ultimate outcome. This is especially true in the case of the individual cell where a given probability of injury or lethality may require numbers of traversals by sparsely and densely ionizing particles that are in the ratio of 1000:1.

The first formulation of microdosimetry was based on the concept of sites that are regions of specified dimensions in which the energy absorbed from ionizing radiations is considered without regard to its microscopic distribution within a site. This approach, which will be termed *regional* microdosimetry, continues to receive major attention because it involves quantities that can, at least in first approximations, be related to radiation effects and that often are subject to accurate measurement.

An alternative, and more advanced, form which will be termed *structural* microdosimetry was originated by Kellerer. Structural microdosimetry permits a detailed description of the microscopic pattern of energy absorption (also termed *inchoate* distribution) and it is of basic importance because the immediate effect of radiation is essentially determined by the intersection of this pattern and that of sensitive components in irradiated matter. *Migration* of energy between the points of absorption and these components can be a complicating factor that can however -at least in principle- be accommodated in structural microdosimetry.

Like other branches of physics microdosimetry has *experimental* and *theoretical* aspects. The former deals with the measurement of microdosimetric quantities and the latter is concerned with relations between these, and other more general, physical quantities. The experimental approach is especially suited to regional microdosimetry although the results obtained can be analyzed to provide information on structural microdosimetry. Conversely, the theoretical approach is of primary importance in structural microdosimetry but it can be extended to regional microdosimetry.

I.2   The Transfer of Energy from Ionizing Radiation to Matter

In its usual interpretation the term *ionizing radiation* is said to apply to electromagnetic radiation and to atomic fragments (electrons and nuclear constituents) with sufficient energy to directly or indirectly eject electrons from the atoms of irradiated matter. Uncharged particles, such as photons or neutrons, ionize indirectly by producing charged particles, such as electrons or protons, that ionize directly. Although ionizing radiations have the dual nature of waves and particles they should usually be regarded to consist of individual particles when they transfer energy to matter.

The degree of ionization resulting from irradiation can be readily determined in a *gas* where the application of an electric field results in a current of electrons (or negative ions) and positive ions. When β (the ratio of the velocities of a charged particle and of light) exceeds about 10^{-3} roughly half of the energy is expended in collisions resulting in ionization with the remainder appearing in the form of excitations (an increase of electron energy within the parent atom) or as kinetic energy of very low energy electrons that do not ionize in turn (ICRU, 1979). As the velocity of charged particles decreases the probability of ionization is reduced and reaches zero when the energy transferred to atomic electrons is less than of the order of a few eV. The distinction between particles that ionize and those that cause only, generally less effective, excitations depends on the ionization potential of the irradiated gas. The particles emitted by almost any practical source of ionizing radiation have energies of kiloelectron volts or more and the fraction of the energy they dissipate by ionization is largely independent of their type or energy. However, there can be thousandfold differences in the spacing of collisions in charged particle trajectories and also appreciable differences in the degree to which energy is deposited by secondary radiations in branches of the primary particle track.

The simple distinction between ionization and excitation in a gas has led to adoption of the term "ionizing" to radiations interacting with matter in the condensed (solid or liquid) phase where the distinction between these processes may be blurred and the effect under consideration may be different. Ambiguities arise from interatomic energy transport processes, such as those in conduction bands that require minimal energy to permit migration of electrons from their parent atoms. Hence the criterion of electron mobility in a semiconductor represents a lower energy requirement than ionization in a gas.

A more general interpretation of the term "ionizing radiation" will be adopted here. It is based on a concept termed the *relevant (energy) transfer point.*

The physical, chemical, biological (and other) effectiveness of ionizing radiations is due to the fact that it can cause important changes in irradiated matter that are due to single interactions. In a gas the preeminent change is ionization, the mechanism embodied in the adjective "ionizing". In a semiconductor the change may be the raising of an electron into a conduction band and in biology the term change refers to a result of a single interaction between radiation and tissue which can cause or contribute to the causation of an observable effect. According to this view retention of the term "ionization" is similar to the extension of the meaning of "oxidation" in chemistry which refers to electron removal in reactions that may not involve oxygen.

The discrete interactions of ionizing radiation and matter are said to occur at (energy) transfer points with disregard of the quantum mechanical uncertainties regarding to location of the point. It is necessary to distinguish the energy

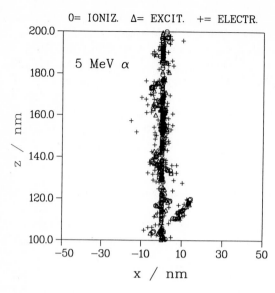

Fig.I.1. A Monte Carlo simulation of energy deposition by a 5-MeV α particle in water. Energy transfer points (ionizations and excitations) produced by this track are indicated with 0 and Δ, respectively. Also shown by + are the positions where secondary electrons (δ rays) stop (i.e. reach thermal energies). Note that this representation is a projection in the (x,z) plane of a 3-dimensional track; therefore the density of transfer points shown here does not correspond to the actual density in the track. However, the α particle travels along the z axis

expended at the transfer point from that *deposited* or *absorbed* there because some energy may be further transported by ionizing radiation such as delta radiation (Fig. I.1).

The energy absorbed at a transfer point is termed an energy deposit, ε_i, and defined by

$$\varepsilon_i = E_{in} - E_{out} + Q \tag{I.1}$$

where E_{in} is the energy (excluding rest energy) of the ionizing particle arriving at the transfer point and E_{out} is the sum of the energies (excluding rest energies) of all ionizing particles leaving the point. The term "ionizing particle" refers to directly as well as to indirectly ionizing radiations. Q is the energy involved in changes of rest mass. In accord with the usual definition, Q is negative in an increase and positive in a decrease of rest masses (see also Eq.III.27). It relates to such changes as nuclear reactions and pair production and is usually equal to zero.

Because of the production of delta rays[1] the majority of deposits by almost any ionizing radiation are produced by electrons and the changes result from their

---

[1] The term "delta ray" means a secondary electron resulting from an ionization process.

interactions with atomic electrons. However, the energy transferred in electronic collisions may approach zero. Hence only a fraction of the energy imparted by ionizing radiation is instrumental in causing changes and any effects resulting from them. The transfer points where this fraction is absorbed will be termed relevant transfer points. As considered in detail in Sect. IV.3.2, in the case of irradiation of a gas the energy expanded in producing the changes (ionizations) at the relevant transfer points is typically one half of the energy absorbed in the gas. Although this fraction may differ for other effects or media, corresponding relations govern any measurements of ionizing radiation as well as any of its effects.

The minimum energy required for the change under consideration will be symbolized by ω and transfer of energy in excess of ω will be termed a *significant energy deposit* at a *significant transfer point*. A significant transfer point becomes therefore *relevant* if a change occurred. When - as in the case of particles of very low energy - $\varepsilon_{i,max} < \omega$ i.e. the largest transfer is *not significant*, no relevant transfer points can be produced. Such particles are *non-ionizing* and their energy is considered locally absorbed rather than included in E_{out}. These postulates constitute a definition of the term "ionizing radiation" and permit specification of the geometric distribution of the energy absorbed from ionizing radiation. Further energy transport occurs by non-ionizing radiation (ultimately by heat conduction). N_R, the average number of *relevant* transfer points, is given by:

$$N_R = \int_{\omega}^{\varepsilon_{i,max}} N(\varepsilon_i)p(\varepsilon_i)d\varepsilon_i \qquad (I.2)$$

where $N(\varepsilon_i)d\varepsilon_i$ is the number of transfer points in the interval $d\varepsilon_i$ and $p(\varepsilon_i)$ is the probability that the change results from a deposit ε_i.

Where Q is zero the production of relevant transfer points requires that E_{in}, the energy of ionizing particles is at least equal to ω. In the case of photons it need not exceed this value. In the case of electrons and especially that of heavier particles the kinematics of interaction requires larger energies. However, for particles that can interact so that $Q > 0$, there may be no lower limit to their kinetic energy (e.g. thermal neutrons).

The term *change* refers to alterations that may be due to different mechanisms. Thus if the change under consideration is ionization in a polyatomic gas it requires energy transfers that vary among atoms and their electronic shells and ω is equal to the lowest ionization potential.

The probability that *changes* result in *radiation effects* depends on the number of relevant transfer points which, on the average, is proportional to the absorbed dose. The effect probability depends on the location of relevant transfer points. When the effect is due to more than one change, the effect probability depends on the geometrical disposition of relevant transfer points. There is almost invariably a statistical correlation between these points in the tracks of charged particles.

The specification of this correlation is the principal objective of microdosimetry which employs the basic concept of the *(energy deposition) event*. An event is the production of statistically correlated transfer points. Examples are the track of an energetic charged particle and/or its secondary electrons, or the charged particles produced in a nuclear reaction. The energy deposited in an event, ε, is the sum of all energy deposits and in regional microdosimetry one considers the energy deposited in a delimited region termed the *site*.

From the considerations presented above, a quantity more pertinent than the energy deposited in an event is the number of relevant transfer points. Although these quantities are, on average, proportional to each other this proportionality cannot be assumed when the energy is small. Thus a charged particle may traverse a proportional counter employed in microdosimetry and deposit energy but produce no relevant transfer points because excitations but no ionizations occur.

Corresponding conditions exist with respect to other changes and in particular those involved in radiobiology. Hence a basic consideration concerns the relative numbers of relevant transfer points in the gas of microdosimetric counters and in biological systems. This subject is considered in Sect. IV.3.2.

I.3 Stochastic Quantities

The discreteness of matter (which consists of atoms) and of radiation (which consists of particles) causes fluctuations in many physical processes. Examples of fluctuating quantities are the collective mass of molecules in a volume in a gas of given density, or the number of electrons flowing during a time interval in an electric current. Frequently the departures from mean values can be ignored because they are negligible. However in the case of ionizing radiation, where a single particle can kill a cell and a single ionization can decompose a critical molecule, statistical fluctuations are important and they may be a major source of uncertainty of radiation action.

In common with other physical phenomena, radiation and its interactions can be numerically specified by pairs of quantities that have the same physical dimension. For example, the quotient of the number of disintegrations in a sample of radioactive material and the time in which they occur is a quantity that differs from the activity which is the mean number of disintegrations per unit of time. The former is a *stochastic* quantity; the latter is a *non-stochastic* quantity. These are two different quantities and they usually do not have the same value. In successive equal time intervals the number of disintegrations observed may decrease or increase while the activity only decreases continuously according to the decay constant. Furthermore the activity refers to a given instant of time while the number of disintegrations refers to a time interval.

A plot of the corresponding integral distributions, i.e. of N, the number of particles emitted during a period of time, t, consists of randomly spaced discontinuous steps. The temporal rate of emission at t is either zero or infinite and the differentiation required to determine dN/dt can only be performed on a smoothed curve of N versus t. This yields the non-stochastic activity that is, strictly speaking, unmeasurable and therefore an abstraction.

What, in fact, is always measured is the stochastic quantity which can, in principle, be determined with arbitrary precision and refers to an interval that needs to be specified. The values of the stochastic quantity are subject to a probability distribution[2] which - in the case of radioactive decay discussed here - is Poissonian. Thus, with an activity, A, the probability distribution of the number of disintegrations, N, in a time interval τ is given by

$$P(N; A\tau) = e^{-A\tau} \frac{(A\tau)^N}{N!} \tag{I.3}$$

Although the shape of the distribution is known, the necessarily limited number of measurements of unpredictable values makes it impossible to determine its mean value, A, with absolute precision. When the average number of disintegrations during the time interval, τ, is large the distribution of N is narrow and a single measurement may provide an adequate estimate of A although there remains a finite probability that N is equal to zero.

Nuclear disintegration is only one of many random processes in radiation physics. Thus, if an object is exposed to a radioactive source emitting gamma radiation, the energy received in a volume within the object depends not only on the fluctuating number of photons emitted by the source but also on a series of other stochastic processes. These include absorption of photons, scattering of photons, liberation of electrons, attenuation of electrons, the rate of energy loss of primary electrons and production of the several generations of the delta radiation (i.e. the secondary, and higher order electrons) produced by them. Calculations are frequently concerned with the average or expectation values of the quantities involved.

---

[2] The abbreviated form of the more accurate term *probability density distribution* is employed when there should be no confusion with the meaning of *probability distribution* as applied in statistics where it is an integral of the probability density distribution.

By employing these non-stochastic quantities it is possible to derive the expected energy absorbed per unit mass at a point in the object which is the *absorbed dose*, D. In general D varies in space and time, but because it is a non-stochastic quantity, it can be defined as a point function in differential form:

$$D = \frac{d\bar{\varepsilon}}{dm} \tag{I.4}$$

where $d\bar{\varepsilon}$ is the mean energy absorbed in dm; and its temporal variations is expressed by the *absorbed dose rate*, $\dot{D}$, as:

$$\dot{D} = \frac{dD}{dt} = \frac{d^2\bar{\varepsilon}}{dm\,dt} \tag{I.5}$$

In a volume V surrounding the point at which the absorbed dose is equal to D the corresponding stochastic quantity *specific energy*, z, is defined by

$$z = \frac{\varepsilon}{\rho V} = \frac{\varepsilon}{m}, \tag{I.6}$$

where ρ is the density. $\overline{D}$, the mean value of the absorbed dose D in V is equal to $\bar{z}$, the mean value of the probability distribution f(z) in V. With decreasing V the variance of this distribution increases and there is an increasing probability that z is equal to zero because V may not contain even one energy transfer point when the absorbed dose is low. Statements on z and f(z) must therefore include information on V. The latter is frequently taken to be spherical and its diameter, d, is then sufficient for specification. If V is non-spherical, more parameters, and even its orientation with respect to the radiation field are, at least in principle, required. At large absorbed doses f(z) becomes a bell-shaped curve about the mean value D.

Microdosimetry is usually concerned with volumes that are sufficiently small so that in most types of irradiation the dose can be considered to be constant. In this case $\overline{D} = D$ at any point in the volume and $\bar{z} = D$. This will be generally assumed to be the case. However the precise definition of D is

$$D = \lim_{m \to 0} \bar{z} \tag{I.7}$$

Although an absorbed dose D cannot be exactly determined it can be evaluated with increasing precision in a multiplicity of measurements in which the mass m is repeatedly exposed. In a conceptual, rather than practical, alternate procedure large numbers to the same radiation fluence masses m are simultaneously exposed to the same radiation fluence.

The mean energy, $\bar{\varepsilon}$, is deposited in events, the number of which conforms with Poissonian statistics. Because the individual energy absorbed per event is variable (see Chap. II) this results in a more complex *compound Poisson* process. Hence, while as shown in Eq(I.3) knowledge of A is sufficient to determine the distribution of N for any τ, knowledge of D is not sufficient to determine f(z). This is the principal reason why specific energy and certain other microdosimetric quantities are named and defined separately from the corresponding non-stochastic quantities.

In usual measurements of absorbed dose it is unnecessary to investigate f(z) when D is measured because a large number of events occur in the dosimeter. What is more important is the converse objective of obtaining f(z) for a given value of D, especially when the number of events, is small. We shall emphasize this by using in the following the more explicit notation f(z;D) instead of f(z).

If the specific energy in a critical volume can be correlated to a biological or other effect, the *effect probability* can be denoted by $\eta(z)$. It is rarely possible or of interest to determine the effect of a known specific energy in a unique site. Thus, in the typical case of cell injury a large number of cells is exposed to an absorbed dose D. The probability, $\eta(D)$ that a cell is affected is then

$$\eta(D) = \int_0^\infty \eta(z) f(z; D) \, dz \qquad (I.8)$$

$\eta(D)$ can be estimated as the fraction of cells affected. In general the estimation of $\eta(z)$ requires knowledge of f(z;D).

In the special case when $\eta(z)$ is proportional to z, with a proportionality constant K,

$$\eta(D) = K \int_0^\infty z f(z; D) \, dz = K\bar{z} = KD, \qquad (I.9)$$

the shape of f(z;D) is irrelevant and the effect probability is equal for equal absorbed doses of different radiations. The fact that this is rarely the case means that $\eta(z)$ is not proportional to z and this is the fundamental reason why f(z;D) is needed in quantitative treatments of biological and other effects of ionizing radiation that are based on specific energy.

The *lineal energy*, y, is defined by

$$y = \frac{\varepsilon_1}{\bar{l}} \qquad (I.10)$$

where $\bar{l}$ is the mean diameter of a volume and ε_1 is the energy imparted in a *single* event. If the volume is taken to be spherical then the specification of y in terms of the diameter d is given by the relation $\bar{l} = 2d/3$, a result obtainable from Cauchy's theorem (Cauchy, 1908).

While the specific energy is the stochastic analog of the absorbed dose, the relation between the lineal energy and the somewhat similar non-stochastic quantity *linear energy transfer* (LET) is not a simple analogy. LET defined as dE/dx represents the mean energy, dE, lost in electronic collision by a charged particle traversing a distance dx and thus includes energy transferred to delta radiation. The restricted linear energy transfer, L_Δ, excludes *kinetic* energies of these electrons beyond a cut-off Δ. The lineal energy is subject to a *geometric* cut-off because ε_1 of Eq(I.10) is the sum of the energy deposits in the volume. These may be due to various statistically correlated particles involved in an event. When d is large some of the particles may not entirely traverse the volume, and when d is small the concept of a particle track may become vague; in contrast, there is no ambiguity in the meaning of *event*.

In an irradiated medium the lineal energy y is subject to a probability density distribution, f(y). Information on distributions of microdosimetric quantities and their interrelations is given in Chap. II.

Although stochastic quantities are of especial importance in microdosimetry, non-stochastic ones are useful as well. They include various moments of the stochastic distributions. The *event frequency*, Φ^* (the mean number of events in a site per unit of absorbed dose) is the reciprocal of z_F, the *average specific energy absorbed in single events* (see Chap. II). The *proximity function*, t(x), which is a key function of structural microdosimetry refers to the distribution of distances between pairs of energy deposits in irradiated matter.

I.4 Spatial Aspects of Microdosimetry

Although there is no inherent constraint on the spatial scale of microdosimetry there are limitations on the domains in which it can be useful.

The most obvious of these relates to the dimensions of the volumes in the irradiated material in which the specific energy, z, can differ appreciably from its average value, the absorbed dose, D. z is subject to three sources of fluctuations, i.e., variations in the number of events in the site, in the number of collisions occurring in these events and in the energy locally deposited in individual collisions. These fluctuations become more pronounced as the site diameter is reduced; ultimately there is an increasing probability that no event occurs in a sufficiently small site. When the size of sites is such that, on average, no energy is deposited in half of them the mean specific energy in the other half is equal to 2D. If this is taken to be the criterion of significant average deviation of z from its mean value, the curves in Fig.I.2 give the *maximum* diameters of spherical sites (in soft biological tissues) in which microdosimetry is important for photons and neutrons of greatly different energies. Their magnitude evidently depends on D. Graphs corresponding to other criteria can be constructed on the basis of information given in Chap. V.[3]

While information of the type shown in Fig.I.2 can be used to delineate the scale where microdosimetry may need to be considered, more specific considerations relate to the dimensions of the domains in which the local energy concentration determines radiation effects. These vary depending on the kind of effect. To illustrate this, in the following the particularly important subject of the action of ionizing radiation on the cells of higher organisms will be considered.

A fundamental concept in radiobiology is that of the *relative biological effectiveness* (RBE) of radiation which is defined by

$$RBE(S) = \frac{D_L}{D_H} \tag{I.11}$$

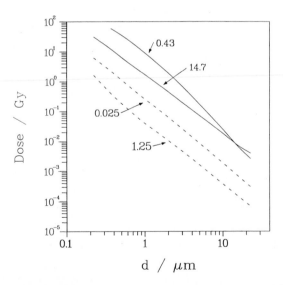

Fig.I.2. Diameters, d, of sites in which at an absorbed dose, D, there are no events in one half of the sites and the mean specific energy, $\bar{z} = 2D$, in the other half. Solid lines: 0.43 and 14.7 MeV neutrons; dashed lines: 0.025 and 1.25 MeV photons

---

[3] Small events due to delta rays injected into the site are not included.

D_H is the absorbed dose of radiation, H, at which the probability of a given biological effect is equal to that at an absorbed dose D_L of a reference radiation, L. The reference radiation is usually orthovoltage (e.g. 250 kVp) x radiation or hard (e.g. ^{60}Co) gamma radiation.

Differences in the biological effectiveness of radiation are attributed to differences in the energy absorbed in sensitive regions of cells and there is an obvious question as to the *dimensions* of the sites in which the energy concentration is *critical*. It seems rather well established (see Chap. VI) that there are at least *two* domains of importance.

The cell nucleus is evidently a principal target for radiation injury. Its dimensions are of the order of *micrometers*. The importance of the cell nucleus is very likely due to the fact that it contains virtually all the DNA in the cell. The general shape of the DNA molecule has been approximated by a cylinder having a diameter of about two nanometers. Various degrees of folding are known to exist and intermediate radiation products (such as active radicals) can diffuse over comparable distances. Nevertheless, energy concentration in volumes having dimensions of less than *ten nanometers* should be of critical significance in DNA damage. As shown in Chap. V such changes can only be produced in single events, at all but extremely high absorbed doses.

The kinetic energy expended by charged particles in such volumes is usually not absorbed within them (most of it being removed by δ rays leaving the volume) and the concept of the particle track needs to be replaced by that of isolated *relevant energy transfer points* that are produced by one or a few delta rays, or by the primary particle. Because of the large fluctuations involved, the concept of LET (i.e. the mean energy *lost* per unit distance in the track) is particularly inappropriate for numerical assessments of DNA damage and microdosimetric distributions in nanometer domains become necessary.

It is generally believed that single strand breaks (SSB) are readily repaired but that their pairwise combination in double strand breaks (DSB) can cause simple chromosome breaks that can subsequently result in deleterious effects on cells, including lethality. In further dual interactions, broken chromosomes can form two-break aberrations such as dicentric or ring chromosomes that are especially deleterious. As discussed in detail in Chap. VI such lesions are generally the product of energy depositions that are separated typically by several tenths of a *micrometer*. This defines therefore the second domain of importance in microdosimetry. Because the interaction probabilities may depend on intracellular distances one is considering "average" site dimensions. *Regional* microdosimetry and its physical quantities that relate to the site concept is largely focussed on this scale. Because of the limited range of delta rays the kinetic energy expended by a charged particle in volumes of this (i.e. micrometer) magnitude is usually largely absorbed in them. Calculations of energy deposition in the site frequently ignore the small difference between primary particles and their δ rays, and many measurements as well as calculations do not determine the corresponding events due to delta rays that are injected by primary particles passing just outside the boundary of the site. Although the energy deposited by the delta rays is usually relatively small their number can be appreciable, e.g. when a 5 MeV α particle traverses 1 μm sites (see Fig.IV.6).

If the chromosomal breaks are produced in the same event the yield of resulting dicentric aberrations, to take an example, is proportional to the absorbed dose. However, if the two breaks are produced in different events the average yield of these aberrations increases as the square of the absorbed dose and this mechanism becomes therefore increasingly important at large absorbed doses, especially of photons. These interactions are of primary importance in the majority of radiobiological experiments with low-LET radiation and they also contribute principally to the results of conventional radiotherapy irradiations.

At distances that are less than the diameter of the DNA helix, atomic dimensions are approached where quantum mechanical constraints obscure even the precise location where charged particles transfer energy. However, at that level one is dealing with energy absorbed at single transfer points which is essentially the same for all ionizing radiations. Such domains are too *small* for microdosimetry because it is concerned with the spatial aggregation of transfer points that is responsible for the effects of ionizing radiation on molecular and larger structures.

I.5 Temporal Aspects of Microdosimetry

Considerations regarding temporal aspects of microdosimetry are primarily based on the fact that ionizing radiation imparts energy to matter in discrete events. With the definition of events as statistically correlated appearances of energy deposits (Sect. I.2) it can usually be assumed that these occur within a time period of less than a picosecond because they are produced by swiftly moving charged particles in microscopic volumes. An exception that is rarely important is the generating of a radioactive atom by one particle and the subsequent decay of that atom accompanied by the emission of another ionizing particle. In this case the two particles may have to be considered to produce separate events if their appearances are separated by a time interval that is sufficient to alter their combined effect.

Although the term *event* does not necessarily refer to energy deposition within a restricted volume, this is nevertheless often implied and, in particular, the event frequency Φ^* applies to what are usually assumed to be spherical regions (sites) of a specified diameter.

It is a consequence of the concept of relevant transfer points that *events* are *relevant* only when they produce at least one relevant transfer point. For instance, in experimental microdosimetry only the events containing transfer points relevant to ionization are registered.

Effects of radiation due to single events occur at a rate that is proportional to the absorbed dose. In the example used in the previous section (dicentric chromosomal aberrations resulting from pairs of single chromosome breaks) the yield of aberrations induced by single events is given by αD (α is a constant) and is expected to be unaffected by dose protraction (reduction of dose rate or fractionation). As discussed in detail in Sect. VI.1 many of the cellular effects of ionizing radiation are due to more than one of the changes that occur at relevant transfer points. For densely ionizing radiation (high LET) the required number of changes may be produced by a single particle. For all but very small doses of low-LET particles, however, it is more probable that these changes result each from different events. For this latter situation, the yield of dicentric aberrations depends quadratically on dose, that is as βD^2, the exponent of 2 being the result of two chromosome breaks (again, each produced by a separate event) contributing to the dicentric formation.

Microdosimetry sets rigorous limits on the probability of having, at a given dose D and for a given sensitive volume V, two independent events. Thus, by Poissonian statistics the probability of two or more events, P_2 is given by:

$$P_2 = 1 - e^{-\phi^* D}[1 + \phi^* D] \qquad (I.12)$$

and because the event frequency, Φ^*, depends on both V and the kind of radiation one can evaluate P_2 in a straightforward manner. The (not unexpected) result is that this quantity increases with V and decreases with LET of the radiation. At the nanometer domain, it is quite unlikely that DNA double strand breaks (the precursors of simple chromosomal breaks) would be affected by the protraction of the irradiation for any of the doses commonly used in radiobiology (tens of Gy or less). For instance, for a spherical volume with d=10 nm a radiation depositing on average 10 keV/μm has $\Phi^* = 0.005$/Gy which means that at a dose of 200 Gy one has, on average, one event.

As discussed in detail in Sect. VI.1 many if not most cellular effects resulting from other than very small doses of low-LET radiation are caused by two events which necessarily occur at different times and in protracted irradiation the damage produced by the first event may be repaired before it can interact with that produced in the second event. Microdosimetric considerations indicate that the events must occur in volumes having dimensions of the order of one micrometer in which the damaged entities must interact and there is a possibility that even with instantaneous irradiation repair occurs before the interaction can take place. Experimental studies in which the rate of recovery of irradiated cells is equated with the rate of DNA repair generally fail to recognize this possibility.

Because of the incoherence (statistical independence) of events they occur at random times when a given absorbed dose, D, is applied during a time, T. When protraction is achieved by continuous irradiation during T the dose rate is =D/T. This can be considered to be the limiting case of the application of f fractions in which absorbed doses D/f are instantaneously applied separated by T/(f-1) intervals. Repair results in an effective absorbed dose $[q(f)]^{1/2}D$ which, in the case of continuous irradiation is symbolized by $[q(\infty)]^{1/2}D$.

In a simplified illustration one may consider the reduction of the effectiveness of two events separated by time T assuming that repair eliminates damage according to exp(-r) where $r=T/t_0$ with t_0 being a characteristic time required for repair. It is also assumed that there is no change of radiation sensitivity during T.

On the basis of a general treatment of the temporal dependence of radiation sensitivity (Kellerer and Rossi, 1972) it can be shown that

$$q(f) = \frac{1}{f} + \frac{2}{f^2} \sum_{i=1}^{f-1} (f - i)e^{-\frac{ir}{f-1}}$$ (I.13)

with $f = \infty$ this reduces to

$$q(\infty) = \frac{2}{r} - \left(\frac{2}{r^2}\right)(1 - e^{-r})$$ (I.14)

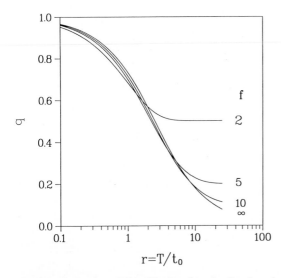

Fig.I.3. Reduction of the effective dose by fractionation according to a simple model. The reduction factor, q, is given versus r, the total period of irradiation expressed in multiples of the characteristic recovery time t_0 for exponential repair. Number of fractions are: 2, 5 and 10. $f = \infty$ represents continuous irradiation at a dose rate D/T

Figure I.3 shows q(f) for f = 2, 5, 10 and ∞. At low values of r the splitting of D into a few fractions results in a moderately stronger reduction. At large values of r only the dose received in the last fraction is significant. When a reduction of no more than 90% is involved 10 fractions are nearly equivalent to continuous irradiation.

Chapter II
Microdosimetric Quantities and their Moments

II.1 Definitions

The principal microdosimetric quantities: *specific energy*, z, and *lineal energy*, y, have been referred to in Sect. I.3. The formal definitions by the International Commission on Radiation Units and Measurements (ICRU, 1980) are:

The *specific energy* (imparted), z, is the quotient of ε by m, where ε is the *energy imparted* by ionizing radiation to matter of mass m:

$$z = \frac{\varepsilon}{m} \qquad (\text{II}.1)$$

Unit: J kg^{-1}. The special name for the unit of specific energy is *gray* (Gy):
1 Gy = 1 J/Kg

The *lineal energy*, y, is the quotient of ε by $\bar{l}$, where ε is the energy imparted to the matter in a volume of interest by an energy deposition event, and $\bar{l}$ is the *mean chord length* in that volume:

$$y = \frac{\varepsilon}{\bar{l}} \qquad (\text{II}.2)$$

Unit: J m^{-1}. ε may be expressed in eV and hence y may be expressed in eV/m or some convenient submultiple or multiple such as keV/μm.

Note that in these definitions ε refers to energy imparted in a *single event* in the case of y, but in any number of events in the case of z.

II.2 Microdosimetric Distributions and their Moments

In the definitions of y and z, $\varepsilon = \Sigma\varepsilon_i$ is the sum of a finite number of energy deposits, ε_i. Unless $f(\varepsilon_i)$, the probability distribution of ε_i is continuous microdosimetric spectra must be discontinuous. In practice these spectra are measured or calculated on the basis of a limited number of discrete values.

In measurements employing tissue equivalent gases additional coarseness is due to the fact that only the number of *relevant transfer points* (i.e. ionizations, see Sect. I.2) is determined. When ε is sufficiently large this approximation may be sufficient. However, even in 1 µm sites some radiations (e.g. relativistic electrons) cause, on the average, only a few ionizations. The quasi-continuous pulse height spectrum recorded in such instances is merely due to technical limitations (e.g. the statistics of avalanche formation in the counter) and it is possible to de-convolute such spectra to determine the number of ionizations (see Sect. IV.3.4).

The main reason for considering y and z to be continuous random variable is mathematical convenience; this assumption is adopted in the following.

The probability that the lineal energy produced by an *event* is in the interval [y,y+dy] is f(y)dy. Analogously, f(z)dz is the distribution of specific energy produced in multiple events by protracted irradiation.

It is useful to make a distinction between the *single-event* microdosimetric spectrum, $f_1(z)$, and the general (multi-event) distribution, f(z). By definition, $f_1(z)$ measures the probability distribution of z *conditioned* by the fact that one event (and only one) took place; $f_1(z)$ is not defined at z=0.

Averages of f(y) and $f_1(z)$ are the *frequency average*, which is the first moment of these distributions, and the *dose average* which is the ratio of the second and first moments:

$$y_F = \int_0^\infty y f(y)\,dy; \quad y_D = \frac{1}{y_F}\int_0^\infty y^2 f(y)\,dy,$$

$$(\text{II.3})$$

$$z_F = \int_0^\infty z f_1(z)\,dz; \quad z_D = \frac{1}{z_F}\int_0^\infty z_2 f^1(z)\,dz.$$

The first and second moments of f(z) are denoted by $\bar{z}$ and $\overline{z^2}$. The *event frequency*, Φ^*, is defined as

$$\Phi^* = \frac{1}{z_F}.$$

$$(\text{II.4})$$

It represents the average number of microdosimetric events per unit dose. One may derive $f(z)$ from $f_1(z)$ for sites having received an absorbed dose D in an *average number* $\Phi^* D = D/z_F$ of events.

Let $p(v;n)$ be the probability for exactly v events, given that the expected (average) number of events is n :

$$n = \sum_{v=0}^{\infty} vp(v; n) \qquad (II.5)$$

Further, let $f_v(z)$ denote the distribution of z obtained in exactly v events. Clearly, f_v does not depend on n. Since events are *by definition* statistically independent one can obtain $f_v(z)$ iteratively from $f_1(z)$ as follows:

$$f_0(z) = \delta(z)$$
$$f_1(z) = f_1(z)$$
$$f_2(z) = \int_0^{\infty} f_1(z')f_1(z - z')dz'$$
$$........ \qquad (II.6)$$
$$f_v(z) = \int_0^{\infty} f_1(z')f_{v-1}(z - z')dz'$$

The integrals of Eqs(II.6) are known as *convolutions* of the two functions (distributions). Thus, for instance, the probability to deposit z about dz in v events is obtained by considering (integrating over) all processes where (z-z') was deposited in v-1 events and z' in the v-th event, irrespective of the value of z'. The distribution, $f(z)$, becomes

$$f(z; n) = \sum_{v=0}^{\infty} p(v; n)f_v(z), \qquad (II.7)$$

where the notation shows the n-dependence explicitly. This expression is of fundamental importance in understanding both microsimetry *and* its applications to radiobiology. Thus:

a) The dependence of $f(z;n)$ on n (and therefore on absorbed dose, to which n is proportional) is entirely contained in $p(v;n)$, a function independent of z.

b) Given $p(v;n)$, the distribution $f(z;n)$ is determined uniquely by $f_1(z)$, see Eq(II.6). The evaluation of $f(z;n)$ is thus reduced to that of the single-event distribution, $f_1(z)$[4].

---

[4] The computer program KFOLD (Kellerer,1985) can be used for this purpose.

c) It should be possible, at least in principle, to invert Eq(II.7) and obtain $f_1(z)$ when $f(z;n)$ and n are known.

Because of the statistical independence of events, $p(v;n)$ is Poisson-distributed:

$$p(v; n) = e^{-n} \frac{n^v}{v!} \qquad (II.8)$$

A number of important relations between the moments of $f_1(z)$ and $f(z;n)$ follow:

Let $\Phi(\omega;n)$ and $\Phi_1(\omega)$ be the Fourier transforms (FT) of $f(z;n)$ and $f_1(z)$, respectively, e.g

$$\phi_1(\omega) = \int_{-\infty}^{+\infty} f_1(z) e^{-i2\pi z\omega} dz. \qquad (II.9)$$

Consider first the FT of $f_2(z)$ in Eq(II.6). Using Eq(II.9):

$$\begin{aligned}
\int f_2(z) e^{-i2\pi z\omega} dz &= \int \left[\int f_1(z') f_1(z-z') dz' \right] e^{-i2\pi z\omega} \\
&= \int f_1(z') \left[\int f_1(z-z') e^{-i2\pi z\omega} dz \right] dz' \\
&= \int f_1(z') e^{-i2\pi z'\omega} \Phi_1(\omega) dz' = \Phi_1^2(\omega).
\end{aligned} \qquad (II.10)$$

This result illustrates a general theorem (the so-called *convolution theorem*) according to which: if $g(x)$ has the FT $G(\omega)$ and $h(x)$ has the FT $H(\omega)$ then their convolution

$$\int g(x') h(x-x') dx' \qquad (II.11)$$

has as FT the *product* $G(\omega)H(\omega)$. The demonstration is identical to that given for Eq(II.10). Returning to Eq(II.6) we note that

$$FT[f_v(z)] = \Phi_1^v(\omega) \qquad (II.12)$$

We take now the FT of Eq(II.7) and use the result, Eq(II.12), to obtain:

$$\Phi(\omega; n) = \sum_{v=0}^{\infty} p(v; n) \Phi_1^v(\omega) = \sum_{v=0}^{\infty} e^{-n} \frac{[n\phi_1(\omega)]^v}{v!} = e^{-n} e^{n\Phi_1(\omega)} \qquad (II.13)$$

This relation provides an analytical expression for the solution $f_1(z)$ of Eq(II.7):

$$f_1(z) = \int \left[1 + \frac{1}{n} \log \int f(z'; n) e^{-i2\pi z'\omega} dz' \right] e^{i2\pi z\omega} d\omega \qquad (II.14)$$

Consider now the v-th derivative of $\Phi_1(\omega)$ with respect to ω:

$$\Phi_1^v = \frac{d^v\Phi_1(\omega)}{d\omega^v} \tag{II.15}$$

From Eq(II.9)

$$\Phi_1^{(v)}(0) = (-2\pi i)^v \int z^v f_1(z)dz = (-2\pi i)^v m_v, \tag{II.16}$$

where we have introduced the notation m_v for the v-th moment of $f_1(z)$. Similarly, we shall use M_v to represent the v-th moment of $f(z;n)$. Taking now explicitly the derivatives in Eq(II.13) we obtain:

$$\Phi^{(k+1)} = n\sum_{j=0}^{k} C_k^j \Phi^{(k-j)} \Phi_1^{(j+1)}, \tag{II.17}$$

where

$$C_k^j = \frac{k(k-1)\ldots(k-j+1)}{j!}. \tag{II.18}$$

Since, by definition

$$\Phi_1^{(v)}(0) = (-2\pi i)^v m_v,$$
$$\Phi^{(v)}(0) = (-2\pi i)^v M_v, \tag{II.19}$$

we obtain from Eq(II.7) a general relation between M_k and m_k, namely

$$m_{k+1} = \frac{M_{k+1}}{n} - \sum_{j=0}^{k-1} C_k^j M_{k-j} m_{j+1}, \qquad k = 0,1,2,\ldots \tag{II.20}$$

In terms of the notations introduced in Eq(II.3):

$$z_F = m_1, \qquad z_D = m_2/m_1 \tag{II.21}$$

Two particular cases of Eq(II.20) are:

$$M_1 = \bar{z} = nz_F \tag{II.22}$$

$$M_2 = \bar{z}(\bar{z} + z_D). \tag{II.23}$$

$\bar{z}$ is usually identified with the absorbed dose, D, because microdosimetry deals with very small volumes [see Eq(I.7)]. With this assumption:

$$M_2 = z_D D + D^2,\tag{II.24}$$

where

$$D = \int_0^\infty z f(z; n)dz.\tag{II.25}$$

It is clear from Eq(II.25) that zf(z;n)dz represents the fraction of dose delivered in the interval dz centered at z, and therefore that

$$d(z; n) = \frac{z f(z; n)}{D}\tag{II.26}$$

is the *normalized* dose distribution in z. Similarly, one can define the single-event dose distribution

$$d_1(z) = \frac{z f_1(z)}{z_F},\tag{II.27}$$

the first moment of which is z_D.

It is customary to represent single-event distributions as a function of lineal energy, y, rather than specific energy. The relationship between z (in a single event) and y is:

$$z = \frac{4y}{\rho S}\tag{II.28}$$

This follows from the fact that, according to a theorem by Cauchy, the mean chord length in a convex site is given by (4V)/S, where V and S are, respectively, the volume and the surface of the site. In the case of usually considered spherical sites $S = 4\pi r^2$ and

$$z = \frac{y}{\pi r^2 \rho}\tag{II.29}$$

In most applications of microdosimetry ρ is taken to be equal to 1 g/cm^3, z is given in the unit gray (Gy=J/Kg), y is expressed in keV/μm and r is in μm. With these units:

$$z = \frac{0.204 y}{(2r)^2}\tag{II.30}$$

Experimental data are usually expressed as a function of y.

II.3 Representations of Microdosimetric Distributions

A problem often encountered when microdosimetric spectra are presented graphically is what representation to use. In the following advantages and disadvantages of various possible graphical representations are discussed and illustrated with microdosimetric spectra for 1-MeV electrons depositing energy in a 1-μm sphere. The spectra shown below have been obtained from charged particle transport calculations in water, as detailed in Sect. V.2; and analyzed with methodology described in Sect. V.3.

A general feature of most microdosimetric distributions is the fact that both the lineal energy, y, and its distribution, f(y), span a rather large spectrum of values. It is not atypical, for instance, that f(y) will take values that range over 8 orders of magnitude. Because of this situation a *linear representation*, f(y) vs y, is rarely employed. The reason for this is clear from Fig.II.1 where, although not visible, y ranges from 0 to about 4 keV/μm. The same spectrum is shown in a *log-log representation* in Fig.II.2, with all the details now clearly displayed. One generally associates with radiation depositing a given value of y a certain biological effectiveness (e.g., Sect. VI.1.4). One of the goals of displaying spectra graphically is to be able to estimate from the plot the fraction of events that have lineal energy values in a given range of interest. Because Fig.II.2 does *not* provide directly this kind of information one resorts to a *yf(y)* vs *log(y)* representation. The basis of this is the fact that:

$$\int_{y_1}^{y_2} f(y)dy = \int_{y_1}^{y_2} [y f(y)] d \log(y).$$
(II.31)

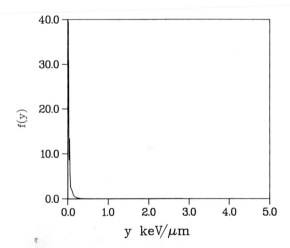

Fig.II.1. Calculated microdosimetric spectrum, f(y), for energy deposited by 1-MeV electrons in a 1-μm diameter spherical volume. In this *linear* representation the details of the distribution are not visible.

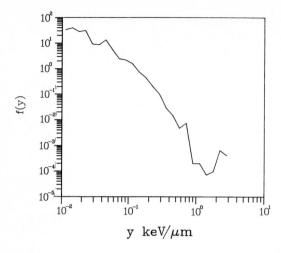

Fig.II.2. The same distribution as in Fig.II.1 but in a *log-log* representation. For this particular spectrum the frequency distribution, f(y), spans 7 decades.

and therefore here the *area* delimited by any two values of y is proportional to the fractional number of events that have lineal energy in that range of y values. The same spectrum in this new representation is shown in Fig.II.3; the details are no longer obscured by the linear representation of the ordinate.

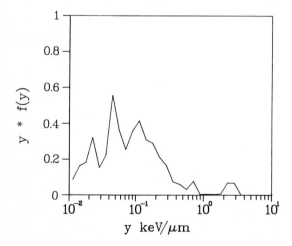

Fig.II.3. The same spectrum as in Fig.II.1 but in a *semi-log* representation. Note that the ordinate has been multiplied by y; therefore, the area under the curve delimited by any two values of y is proportional to the fraction of *events* in this range of lineal energy values. This is the most commonly used representation for the *frequency* of events.

If one is interested in the *dose distribution in y* [proportional to $yf(y)$, see Eq(II.27)] the representation having the same property as that of Fig.II.3 becomes: $y^2f(y)$ vs $\log(y)$. The explanation is similar to that given mathematically in Eq(II.31). Fig.II.4 shows the dose distribution in y; as in Fig.II.3, the area between any two values of y indicates the *dose* delivered in that range. By comparing Figs.II.3 and II.4 it becomes apparent that for this radiation (and in this size volume) most microsimetric events have y values around 0.1 keV/μm, while approximately half of the dose is delivered by events with lineal energy at higher y values (between 2 and 3 keV/μm).

The moments of the microsimetric spectrum can be obtained in either representation as follows (notice the normalization factors):

$$y_F = \int_0^\infty y f(y)\,dy = \frac{\int_0^\infty y^2 f(y)\,d\log(y)}{\int_0^\infty y f(y)\,d\log(y)}$$

$$y_D = \frac{\int_0^\infty y^2 f(y)\,dy}{\int_0^\infty y f(y)\,dy} = \frac{\int_0^\infty y^3 f(y)\,d\log(y)}{\int_0^\infty y^2 f(y)\,d\log(y)}$$

(II.32)

For the spectra shown in Figs.II.1 to II.4: y_F=0.18 keV/μm and y_D=1.18 keV/μm.

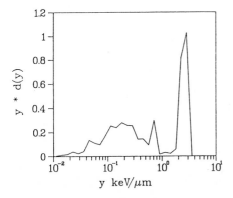

Fig.II.4. The dose distribution, d(y), for the spectrum of Fig.II.1. As in Fig.II.3, the ordinate is multiplied by y and in this *semi-log* representation the area under the curve delimited by any two values of y is proportional to the fraction of *dose* delivered by events with lineal energies in this range. This is the *standard* representation of a microsimetric spectrum.

II.4 Experimental versus Calculated Microdosimetric Distributions

In Sect. II.2 it was pointed out that, with currently available microdosimetric detectors, a measurement consists of recording a spectrum of *discrete* numbers, namely the number of *ionizations* deposited in the sensitive volume. To convert to *energy absorbed* one multiplies this number with the *average energy per ion pair* (W value) for the radiation in question. Figs.II.5 and II.6 show (again for the 1-MeV electrons in 1-μm spheres) the spectra for *ionizations* only. The number of ionizations takes discrete values (1,2,...), and its distribution appears as vertical bars in the figures. In fact, although f(n) is the spectrum *measured* by the counter, this is *not* the spectrum recorded after the experimental apparatus; following energy deposition in the counter volume (ionizations) the charged produced is subject to a process of multiplication (typically by a factor of 100 to 1000) thus making the signal detectable above the electronic noise. As indicated in Chap. IV, where experimental microdosimetry is described in full detail, the multiplication process is statistical in nature and thus one ion pair may result in a spectrum of

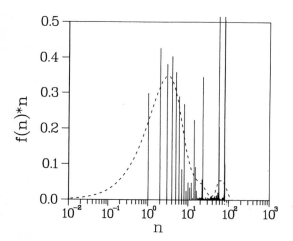

Fig.II.5. A measured microdosimetric spectrum typically records only ionizations. This is indicated by the vertical bars that correspond to events with n=1,2,... ionizations produced by 1-MeV electrons. In a proportional counter the multiplication process leads to a distribution of charges for each ion pair produced in the counter. The net "smoothed" result is indicated by the dashed line. It is this latter that is experimentally recorded.

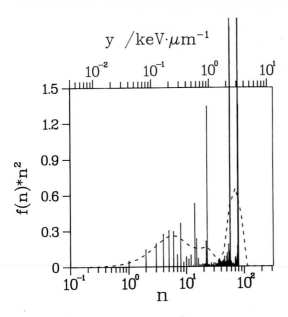

Fig.II.6 Same as Fig.II.5 but in a $n^2f(n)$ versus log(n) representation.

(multiple) charges distributed as[5], for instance, exp(-n). The net result, indicated with dashed lines in Figs.II.5 and II.6, is a process of "smoothing" of the bar spectrum. This may serve as (partial) justification for replacing y - essentially a discrete quantity - by a continuous variable. The smoothing due to multiplication process does not change significantly the moments n_F and n_D of the discrete spectrum. However, a spectrum that includes *all* energy transfers (ionizations and excitations), as displayed in Figs.II.1 to II.4, may have moments that are different from the "ionization-only" distribution. For instance, for these latter one has y_F = 0.28 keV/μm and y_D = 1.62 keV/μm, both significantly larger than the values quoted in the previous Section. This points out to the importance of the concept of *relevant transfer points*, introduced in Sect. I.2.

---

[5] When multiple ion pairs are produced the resulting spectrum is a convolution of the distribution corresponding to one pair.

Chapter III
Interactions of Particles with Matter

III.1 Overview

As already described, regional microdosimetry is concerned with the stochastics of energy *deposition* in *specified* sites. Once the target is defined as a physical ensemble (composition, geometry, etc) the problem is reduced to studying the probability of interaction of particles with the target system - an exercise in quantum mechanics.

The dominant energy deposition process of concern here is the Coulomb interaction between the incident particle and atomic electrons. The *projectiles* are then electrons or heavier charged particles. It is useful for this discussion to make a distinction between *sources* of projectiles (and the cross sections by which the projectiles are generated) and the *projectiles* themselves and their specific energy-transfer interactions. Thus sources include photons (which, for instance, generate energetic electrons via Compton scattering) and neutrons which produce protons and other nuclear recoils via nuclear reactions. Another important source of energetic secondary electrons are, of course, the charged particles themselves. As will be seen further on, this division into sources and charged projectiles has also a practical significance: once the microdosimetric spectra for projectiles are known one can obtain the spectrum corresponding to any source distribution of projectiles (e.g., as present in a neutron field) by a straightforward superposition procedure (see Sect. V.3.3).

This chapter is organized as follows: In Sect. III.2 we collect a number of definitions of quantities and terms frequently used in describing interactions. Sect. III.3 reviews collision kinematics. Cross sections for the production of charged particles by other particles (photons, neutrons and charged particles) are described in Sect. III.4. Next (Sect. III.5), we discuss in detail processes which lead to energy deposition via Coulomb interaction. In electromagnetic interactions projectiles can be the source of other projectiles in the so-called *hard* collisions, as distinguished from *soft* or glancing collisions. In Sect. III.4 we treat those interactions where the kinetic energies of both the incident particle *and* the outgoing electron are considerably larger than the binding energies of the atomic electrons. These interactions - which can be described exactly - are generally rare events. Most of the time the interaction process results in very low energy electrons or excitations; these lead to local energy deposition and, as such, are responsible for most of the microdosimetric phenomena. Unfortunately, the description - theoretical or experimental - of these processes remains tentative.

In the final section we treat phenomena associated with targets of finite thickness: stopping power, range and their statistical fluctuations (straggling).

III.2 Quantities and Terms Relating to the Interaction Between Projectiles and Targets[6]

We are interested in *transport* properties of the radiation and, more specifically, in the probability of a given radiation to undergo an interaction in the medium. The following terminology is used in this respect:

a) The *fluence*, Φ, of particles is

$$\Phi = \frac{dN}{da},\qquad\qquad\text{(III.1)}$$

where dN is the number of particles incident on a sphere of cross section area da. The area must be perpendicular to the direction of the particles and a sphere provides then for this in the simplest manner.

b) The *cross section*, σ, of a target entity for an interaction with incident particles is

$$\sigma = \frac{P}{\Phi}\qquad\qquad\text{(III.2)}$$

where P is the probability of the interaction for one target entity when subject to a fluence Φ. The term *target entity* may refer to atomic constituents, to atoms, or to more complex structures (e.g. biological cells).

In a geometric interpretation σ is the area which each target entity presents to an incident particle for unit probability of interaction. In a uniform, unidirectional

---

[6] The following definitions of radiation quantities are in accord with those in ICRU Report 33 (1980) which should be consulted for further details.

fluence the number of particles traversing a perpendicular plane of unit area is N. In a material layer of thickness dx that contains n targets per unit volume the number of targets per unit area is ndx. The number of interacting particles, dN_i (and hence the number of *affected targets*) is:

$$dN_i = -dN = N\sigma n dx.\qquad\qquad (III.3)$$

(the minus sign is due to the fact that the differential element dN is defined as the difference between the number of particles that did not interact after and before dx). By integration, the fraction of the particles that have not interacted after passing through a layer of thickness x is:

$$\frac{N(x)}{N(0)} = e^{-\sigma n x}\qquad\qquad (III.4)$$

where N(0) is the number of incident particles. This geometric interpretation of σ should not be taken literally; For instance, a quantum-mechanical treatment of the interaction of a particle with a target consisting of a (hard) sphere of radius R shows that for high energy particles the total elastic cross section is $2\pi R^2$ (that is *twice* the classical cross section); while at low energies, the total elastic cross section is $4\pi R^2$, which is *four* times the classical value (Schattschneider, 1986). Generally, σ should be defined as the scattering probability per target and per unit area (ndx is the number of targets per unit area of the layer dx). The dimension of σ is m^2. A more practical unit is the *barn* (1 bn = 10^{-28} m^2). An occasional source of confusion is the omission of the target involved (e.g. electron, atom, molecule) when presenting numerical values for σ.

In transport studies it is necessary to further discriminate among scattering processes by additional variables such as[7] the energy $(\hbar\omega)$ and momentum $(\hbar q)$ transferred by the projectile to the target material; or, equivalently, the energy (E') and scattering angle (θ) of the outgoing particle (Fig.III.1).

---

[7] In line with the general custom, the notations used here, for instance, $\hbar\omega$ and $\hbar q$, referring to energy and momentum, are meant to show explicitly quantum mechanical connections, in this case to frequency (ω) and wave vector (q); and also the fact that in "natural units" ($\hbar =1$, $m_e=1$, e=1, where m_e and e are the electron mass and charge) these two sets of quantities coincide numerically. Natural units, which generally simplify the notations, are being used in the formal derivations given in the Appendix to this chapter. By employing, say, $\hbar\omega$ for the energy transfer - rather than ΔT - the link between expressions given in the main text and in the Appendix is made more clear. When quantum mechanical features need not be invoked, for instance in describing the kinematics of neutron collision, the more conventional notations are being used.

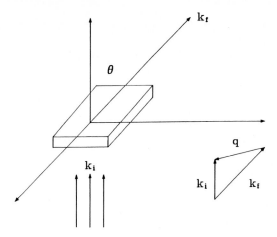

Fig.III.1. Schematic representation of the geometry and notations used in the text to describe particle scattering: Projectiles of momentum $\hbar k_i$ are incident on a target centered at the origin. The outgoing particle shown is emitted at polar angle θ has momentum $\hbar k_f$. $\hbar q$ is the momentum transferred to the target during the scattering event. In most situations it is assumed that the scattering process has cylindrical symmetry, that is it does not depend on the azimuthal angle, ϕ.

Formally, this is expressed in terms of single- or double-differential cross sections. For instance, $[d\sigma/d(\hbar\omega)]d\hbar\omega$ is the probability[8] of energy transfer in the interval $d\hbar\omega$ centered at $\hbar\omega$; and $[d^2\sigma/d(\hbar\omega)d(\hbar q)]d(\hbar\omega)d(\hbar q)$ is the probability of energy transfer in $d\hbar\omega$ and momentum (wave vector) transfer in $d\hbar q$ centered at $\hbar q$. Clearly:

$$\frac{d\sigma}{d(\hbar\omega)} = \int \frac{d^2\sigma}{d(\hbar\omega)d(\hbar q)}\, d(\hbar q), \qquad (\text{III.5})$$

$$\sigma = \int \frac{d\sigma}{d(\hbar\omega)}\, d(\hbar\omega), \qquad (\text{III.6})$$

where the integration limits are set by kinematical constraints (see the next Sect.).

c) The *attenuation coefficient*, μ, for *uncharged* particles (e.g. photons or neutrons) is

$$\mu = -\frac{1}{N_u}\frac{dN_u}{dx}, \qquad (\text{III.7})$$

---

[8] For brevity we shall use "probability" as a substitute for "probability per target and per unit area". The usage should be clear from the context.

where N_u is the number of uncharged particles and $-dN_u/N_u$ is the fraction of uncharged particles that interact in dx. The density-independent *mass attenuation coefficient* is μ/ρ where ρ is the density of the material. By comparison with Eq(III.3)

$$\frac{\mu}{\rho} = \frac{\sigma n}{\rho} = \sigma \frac{N_A}{M_A},$$

(III.8)

where N_A is *Avogadro's number* and M_A is the *molar mass* of the material.

In many instances the principal concern is not with the number of particles and their interactions but with the energy transported and the fraction of this energy transferred to irradiated material.

d) The *mass energy transfer coefficient*, μ_{tr}/ρ, is

$$\frac{\mu_{tr}}{\rho} = \frac{1}{\rho EN}\frac{dE_{tr}}{dx},$$

(III.9)

where dE_{tr}/EN is the fraction of the incident energy of uncharged particles that is transferred to kinetic energy of charged particles in a distance dx.

e) The *mass energy absorption coefficient*, μ_a/ρ, is

$$\frac{\mu_a}{\rho} = \frac{\mu_{tr}}{\rho}(1 - g)$$

(III.10)

where g is the fraction of the energy of secondary charged particles that is lost to brehmsstrahlung in the irradiated material.

f) The *kerma*, K, is

$$K = \Phi_u\left[E_u \frac{\mu_{tr}}{\rho} \right],$$

(III.11)

where Φ_u is the fluence of uncharged particles of energy E_u. K is the energy imparted per unit mass by the charged particles produced by the fluence Φ_u of uncharged particles. The expression in the square brackets of Eq(III.11) is termed the *kerma factor*.

Under conditions of *charged particle equilibrium*[9] the absorbed dose is equal to kerma

$$D = K$$

(III.12)

---

[9] Charged particle equilibrium at a point exists if the charged particle fluence is constant within distances equal to the maximum charged particle range. An equivalent definition is given following Eq(V.76).

g) The *mass stopping power* for charged particles is

$$\frac{S}{\rho} = \frac{1}{\rho}\frac{dE}{dx},$$ (III.13)

where dE is the energy lost by a charged particle in traversing a distance dx.

h) The *linear energy transfer* (LET), L_Δ, is

$$L_\Delta = \left(\frac{dE}{dx}\right)_\Delta,$$ (III.14)

where dE is the energy lost by a charged particle due to collisions with electrons in traversing a distance dx minus the kinetic energy of all the electrons released with kinetic energies in excess of Δ.

L_∞ is equal to S_{col}, the *linear stopping power* due to collisions. It may be replaced by L.

i) The *cema*, C, is

$$C = \frac{\rho}{\Phi_C} S_{col},$$ (III.15)

where Φ_c is the fluence of charged particles other than delta rays. Under the conditions of *delta ray equilibrium* cema is equal to the absorbed dose

$$D = C$$ (III.16)

When in dosimetry calculations the absorbed dose is assumed to be equal to kerma the transport of energy by charged particles liberated by uncharged particles is neglected. In more precise calculations the absorbed dose is assumed to be equal to the cema of these particles; this neglects the generally much less important further transport by the delta rays of the charged particles.

j) The *free path* is the particle trajectory (assumed, for simplicity, to be a straight line) between successive interaction events. Along a free path the energy and momentum of the particle are assumed to be constant. Depending on the interaction process considered, one may have a free path for elastic scattering, ionization, excitation, etc.

k) In an *absorption* event the incident particle disappears (e.g. the photoelectric effect, see Sect. III.4.A).

l) *Scattering* is an interaction event accompanied by energy and momentum transfer to the medium. The projectile re-appears as a scattered particle. Scattering events are classified as elastic or inelastic (see Sect. III.3).

m) M*ean-free path* (mfp). Under the assumption that successive interactions are governed by Poisson statistics, the probability of a free path (no scattering) trajectory s followed by interaction between s and s+ds is

$$f(s)ds = e^{-n\sigma s} n\sigma ds \qquad \text{(III.17)}$$

The *mean free path* is then defined as:

$$\Lambda = \int_0^\infty s f(s) ds = \frac{1}{\sigma n} \qquad \text{(III.18)}$$

Λ^{-1} is sometimes referred to as *macroscopic cross section.* One has:

$$\mu = \Lambda^{-1} \qquad \text{(III.19)}$$

n) The thickness of material required to reduce to half the number of uncharged particles is the *half-value layer*, $d_{1/2}$:

$$d_{1/2} = \frac{\log(2)}{\mu} = \frac{0.693}{\mu} \qquad \text{(III.20)}$$

III.3 Kinematics of the Scattering Process

Kinematics deals with the consequences of energy and momentum conservation in collision processes. Fig.III.2 illustrates schematically a 2-body collision and introduces some of the notations characterizing each particle:

T = kinetic energy
p = momentum
v = velocity
c = velocity of light in vacuo
m = rest mass
$E = T + mc^2$ = total energy
$\beta = v / c$
$\gamma = (1 - \beta^2)^{-1/2}$

Laboratory system (L)

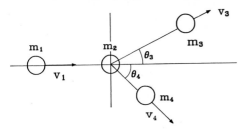

Center of mass system (CM)

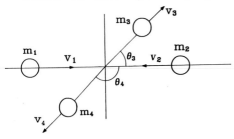

Fig.III.2. In the laboratory system (L) a projectile (mass m_1 and velocity v_1) is incident on a stationary target ($m_2, v_2 = 0$). In this example two outgoing particles, (m_3, v_3) and (m_4, v_4), are emitted at angles θ_3 and θ_4, respectively. When the same two-body scattering is seen from the center-of-mass system (CM) both the projectile and the target move toward each other, see Eqs(III.24,25); and, because of momentum conservation, the outgoing particles are emitted "back to back".

Some of these quantities are related as follows (Jackson, 1975):

$$E = \frac{mc^2}{\sqrt{(1 - \beta^2)}} = mc^2\gamma = \sqrt{m^2c^4 - p^2c^2} \tag{III.21}$$

$$p = \frac{\sqrt{E^2 - m^2c^4}}{c} = \frac{\sqrt{T^2 + 2mc^2T}}{c} = \frac{m\beta}{\sqrt{1 - \beta^2}} = m\beta\gamma \tag{III.22}$$

When $T \ll mc^2$ non-relativistic kinematics may be used.

Many kinematical calculations are simplified if the collision is viewed from the *center-of-mass* (CM) system of reference; this is defined (see Fig.III.2) such that the total momentum of the colliding particles - before or after the collision - is zero. The system of reference at rest relative to the experimenter is termed *laboratory* (L) system; typically, but not always, in L the target is at rest.

Non-relativistically, the velocity of CM relative to L (see Fig.III.2) is

$$v_{CM} = \frac{m_1}{m_1 + m_2} v_1 \qquad \text{(III.23)}$$

Here, the subscripts 1 and 2 refer to the projectile and target, respectively. We shall denote quantities in L by lower-case letters, and quantities in CM by upper-case letters. The velocities in L and CM are obtainable from each other:

$$V_1 = \frac{m_2}{m_1 + m_2} v_1 \qquad \text{(III.24)}$$

$$V_2 = -\frac{m_1}{m_1 + m_2} v_1 . \qquad \text{(III.25)}$$

In the collision of a projectile with a target which is much heavier (i.e. $m_1 \ll m_2$) the distinction between L and CM is unimportant since v_{CM} [see Eq(III.23)] is approximately zero. This will be the case, for instance, when an electron scatters off an atom ($m_1/m_2 \cong 10^{-4}$).

In *elastic collisions*

$$m_1 + m_2 = m_3 + m_4; \qquad T_1 + T_2 = T_3 + T_4 \qquad \text{(III.26)}$$

In *inelastic collisions* the internal energy of the particles involved changes; put differently, there is a transfer of energy between mass and kinetic energy. It follows that

$$m_1 + m_2 = m_3 + m_4 + Q; \qquad T_1 + T_2 = T_3 + T_4 - Q \qquad \text{(III.27)}$$

and the reaction is called *exothermic (endothermic)* if Q>0 (Q<0).

The number of variables (m_i, Q, θ, etc) obtainable from kinematics is equal to the number of equations that express, independently, energy and momentum conservation. For the collision represented in Fig.III.2 we have 3 equations:

$$T_1 = T_3 + T_4 - Q \qquad \text{(III.28)}$$

$$p_1 = p_3 \cos \theta_3 + p_4 \cos \theta_4 \qquad \text{(III.29)}$$

$$0 = p_3 \sin \theta_3 + p_4 \sin \theta_4 . \qquad \text{(III.30)}$$

Consider, as an example, an elastic collision (Q=0) and let p_1 (or $T_1 = p^2_1/2m_1$) and all the masses be known quantities; there are then 4 unknown variables: $p_3, p_4, \theta_3,$

θ_4, which means that once one of these variables is fixed (often θ_3, the angle at which one of the outgoing particles is detected) all other quantities are uniquely determined by Eqs(III.28-30). As for θ_3, to use the same example, its probability distribution, $d\sigma^2/d\theta_3 dT_3$, may be calculated from the interaction potential between projectile and target. Equivalently, one may fix T_3 and consider $d\sigma/dT_3$, but a quantity such as $d\sigma^2/d\theta_3 dT_3$ will not make much sense as θ_3 and T_3 are in a one-to-one relation. If one were to plot $d^2\sigma/d\theta_3 dT_3$ as a function of θ_3 and T_3 the resulting figure will be along a curve representing the function $T_3=T_3(\theta_3)$; this line is sometimes called the *Bethe ridge*. The existence of a Bethe ridge in a two body collision is the result of the target being in a *definite* momentum state. To the extent that the target's momentum may take a series of values with certain probabilities (as will be the case in a more precise, quantum mechanical description of the target) $d^2\sigma/d\theta_3 dT_3$ is not restricted to the Bethe ridge any longer. A situation often encountered is when the momentum transferred in the collision is much larger than the spread in the target momentum; the double differential cross section should then approximate the Bethe-ridge functional form [a pictorial representation of the Bethe ridge may be found in Inokuti (1971)].

The energy transferred in the collision represented by Eqs(III.28-30) is

$$\hbar\omega = \frac{p_4^2}{2m_2},$$ (III.31)

where we take $m_1=m_3$ and $m_2=m_4$ (same particles before and after scattering). The maximum value of $\hbar\omega$ is at $\theta_4=0$, $\theta_3=180°$ (head-on collision):

$$\hbar\omega_{max} = \frac{4m_1 m_2}{(m_1 + m_2)^2} T_1$$ (III.32)

When $m_1 \ll m_2$, $\hbar\omega_{max} = 4(m_1/m_2)T_1$ which means that energy transfer in elastic collisions is exceedingly small. This is the case in electron-atom scattering, and of course more so for electron-molecule or electron-solid scattering.

Finally, in reactions for which $Q<0$, a minimum projectile energy (*threshold energy*) is required for the reaction to proceed. This is given by:

$$T_{threshold} = \frac{-Q\left[-Q + 2(m_1 c^2 + m_2 c^2)\right]}{2m_2 c^2}.$$ (III.33)

Example 1
In the *elastic* collision between a neutron (energy T) and a nucleus of mass number A the energy transferred to the recoiling nucleus is:

$$\Delta T = \frac{2A}{(1 + A)^2} T(1 - \cos\theta_{CM}),$$ (III.34)

where θ_{CM} is the CM scattering angle of the neutron. For θ_L one has

$$\cos \theta_L = \frac{\cos \theta_{CM}}{\left[1 + 2A \cos \theta_{CM} + A^2\right]^{1/2}} \qquad \text{(III.35)}$$

The maximum energy transfer is

$$\Delta T_{max} = \frac{4A}{(1+A)^2} T, \qquad \text{(III.36)}$$

and therefore to moderate (slow down) neutrons light materials (small A) are preferred. Ideally (A=1, hydrogen target) one obtains:

$$\Delta T_{max} = T, \qquad \text{(III.37)}$$

which means that at this (maximum) energy transfer the neutron gives up all its energy. For a hydrogenous target (A=1) $\theta_L = \theta_{CM}/2$. In the CM θ is frequently isotropically distributed. The resulting energy spectrum of recoil particles is shown in Fig.III.3.

Example 2
An electron (momentum $p_i = \hbar k_i$, kinetic energy $T_i = p_i^2/2m$, see Fig.III.1) scatters off a target at an angle θ with respect to the direction of its initial trajectory. At a

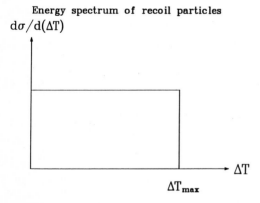

Energy spectrum of recoil particles
$d\sigma/d(\Delta T)$

ΔT

ΔT_{max}

Fig.III.3. If in the CM the angular distribution of recoil particles is isotropic then their energy spectrum in the laboratory system is uniform. This can be seen as follows: The energy spectrum of recoil particles, $d\sigma/d(\Delta T)$, equals (by the chain rule) the product of the angular distribution in the CM, $d\sigma/d(\cos\theta_{CM})$, and $d(\cos\theta_{CM})/d(\Delta T)$. This latter quantity may be estimated from Eqs(III.34,36). One obtains: $d\sigma/d(\Delta T)=[d\sigma/d(\cos\theta_{CM})]2/\Delta T_{max}$. For isotropic distributions (in the CM) the quantity in the square brackets is constant and therefore the laboratory distribution, $d\sigma/d(\Delta T)$, is also constant - as indicated in the figure.

given θ we are interested in the range of energy transfer, $\hbar\omega$, and momentum transfer, $\hbar q$. With $p_f = \hbar k_f$ denoting the final electron momentum we have:

$$\frac{p_i^2}{2m} - \frac{p_f^2}{2m} = \hbar\omega, \tag{III.38}$$

$$\vec{p}_i - \vec{p}_f = \hbar\vec{q} \tag{III.39}$$

From Eq(III.39):

$$p_f^2 - (2p_i \cos\theta)p_f + \left(p_i^2 - (\hbar q)^2\right) = 0, \tag{III.40}$$

and for p_f to take real values it is necessary that

$$(\hbar q)^2 \geq p_i^2 \sin^2\theta, \tag{III.41}$$

The maximum value of $\hbar\omega$ obtains when p_f is minimal [see Eq(III.38)], that is $(p_f)_{min} = p_i\cos\theta$. One obtains:

$$0 \leq \hbar\omega \leq T_i \sin^2\theta. \tag{III.42}$$

Similarly,

$$p_i \cos\theta \leq p_f \leq p_i \tag{III.43}$$

and, since $(\hbar q)^2_{max}$ corresponds to $p_f = p_i$:

$$p_i \sin^2\theta \leq (\hbar q)^2 \leq 2p_i^2(1 - \cos\theta) \tag{III.44}$$

A description of the collision in terms of *energy* and *momentum* transfer is a natural choice in the theoretical treatment of scattering (with the aid of generalized oscillator strength functions, see Sect. III.5.1) but not always in analyzing experimental data, where the variables $\hbar\omega$ and θ may be preferred. This example shows the connection between these two sets of descriptors.

III.4 Sources of Charged Particles

III.4.1 Photon-interaction Cross Sections

Photons may interact with atomic electrons (*photoelectric* effect, *Rayleigh* scattering, *Compton* scattering), nuclei (*photo-nuclear* reactions), the electric field generated by charged particles (*pair production*) and mesons. Of these processes only three (photoelectric, Compton, pair production) are significant in typical microdosimetric applications, basically because of the photon energies (10-20 MeV or less) involved. The essential facts about these processes are:

a) Photoelectric effect (PE) and pair production (PP) are absorption events, i.e. the photon disappears. In Compton scattering (C), as the name implies, the photon undergoes elastic scattering.

b) The dependence of the cross sections, σ, on the atomic number of the target material is: $\sigma_{PE} \sim Z^4 - Z^5$, $\sigma_C \sim Z$, $\sigma_{PP} \sim Z^2$.

c) Photoelectric, Compton and pair production interactions with tissue material (Z<20) are dominant, respectively, in the following energy ranges: <50 keV, (50 keV, 10 MeV), and >10 MeV, see Fig.III.4.

The mass attenuation coefficient for photons is:

$$\mu / \rho = \frac{N_A}{M_A} [\sigma_{PE} + \sigma_C + \sigma_{PP}], \qquad (III.45)$$

[see Eq(III.8)]. For photons the difference between μ/ρ and μ_a/ρ may be quite substantial depending on the photon energy and material. This is illustrated in Fig.III.5.

A. Photoelectric effect

In this process a photon of energy E_γ strikes an electron in an orbit of an atom in the material. The photon is absorbed and the atomic electron ejected with kinetic

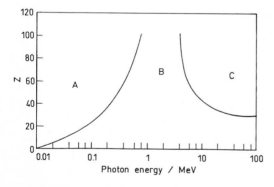

Fig.III.4. The curves delineate the regions where the three types of photon interactions dominate: A Photoelectric, B Compton, C pair production. Z is the atomic number of the absorbing material (after Evans, 1955).

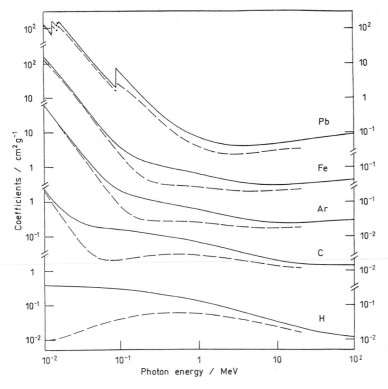

Fig.III.5 Calculated values for the mass-attenuation coefficient (solid line) and mass energy-absorption coefficient (dashed line) as a function of photon energy for several materials as indicated in the figure (after Hubbell, 1982).

energy

$$T_e = E_\gamma - w, \qquad (III.46)$$

where w is the binding energy of the shell from which the electron is removed. To estimate approximately the value of w the following empirical formulae (in units of Ry=13.5 eV) may be used (Fermi, 1950):

$$\begin{aligned} K \quad &shell \quad (Z-1)^2 \\ L \quad &shell \quad (Z-5)^2 \, / \, 4 \qquad\qquad (III.47) \\ M \quad &shell \quad (Z-13)^2 \, / \, 9 \end{aligned}$$

The photoelectric cross section, σ_{PE}, is approximately proportional to $Z^4/E_\gamma^{\,3}$; however, as the energy of the photon increases and approaches the value where the energy balance, Eq(III.46), allows electron removal from an additional (tighter bound) shell there is a characteristic increase (jump) in the cross section termed K, L, M,... edges (see Fig.III.5). This sudden increase in σ is due to additional electrons becoming available and to a general property of this process in that the more tightly bound the electrons are the larger the cross section.

The angular distribution of the ejected electrons (photoelectrons) depends on E_γ. At low photon energies photoelectrons are emitted mostly along the direction of the electrical field vector of the photon, that is at 90° with respect to the incident beam. As E increases the photoelectron is progressively emitted at smaller (more forward) angles.

The vacancy left by the emission of a photoelectron from, say, a K shell is filled by one of the other electrons (L,M,...) and the energy made available by this transition is given off by either x ray emission (*fluorescence*) or by the production of an *Auger* electron from the L or M shells. The ratio between these two processes is called the *fluorescence yield* and increases monotonically with Z.

B. Compton scattering

Compton scattering is an elastic collision between a photon and a free electron at rest. For atomic electrons it is usually assumed that the photon energy is much larger than the binding energy, w, of the atomic electrons and, for any practical purposes, they may be considered free and at rest.

The geometry of the Compton scattering is shown Fig.III.6. The energy of the scattered photon, E_γ', is given by:

$$E_\gamma' = \frac{E_\gamma}{1 + \alpha(1 - \cos\theta)} \tag{III.48}$$

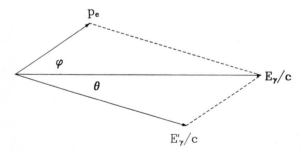

Fig.III.6 In the Compton effect a photon (momentum E_γ/c) scatters off an atomic electron assumed at rest. The electron and photon emerge with momenta p_e and E_γ'/c, respectively.

where

$$\alpha = \frac{E_\gamma}{m_e c^2}.$$ (III.49)

The scattering angle of the electron is given by:

$$\tan \phi = \frac{1}{1 + \alpha} \tan \frac{\theta}{2},$$ (III.50)

and therefore the electron is always scattered forward. The electron energy, T_e, is given by:

$$T_e = E_\gamma - E'_\gamma$$ (III.51)

or

$$T_e = E_\gamma \frac{\alpha(1 - \cos \theta)}{1 + \alpha(1 - \cos \theta)}.$$ (III.52)

The maximum energy a Compton electron may acquire is at the photon angle $\theta = 180°$:

$$(T_e)_{max} = E_\gamma \frac{2\alpha}{1 + 2\alpha}$$ (III.53)

The single differential cross section, $d\sigma/d(\cos\theta)$, for Compton scattering of the photon in the direction θ is given by the *Klein-Nishima* formula (Rossi, 1952):

$$\frac{d\sigma_e}{d(\cos \theta)} = \pi r_e^2 \frac{1 + \cos^2 \theta}{[1 + \alpha(1 - \cos \theta)]^2}$$

$$\left\{ 1 + \frac{\alpha^2(1 - \cos \theta)^2}{(1 + \cos^2 \theta)[1 + \alpha(1 - \cos \theta)]} \right\}$$ (III.54)

where $r_e = e^2/mc^2$ is the *classical electron radius*.

The following equation describes the *total* Compton cross section

$$\sigma_c = 2\pi r_e^2 \left\{ \frac{1 + \alpha}{\alpha^2} \left[\frac{2(1 + \alpha)}{1 + 2\alpha} - \frac{\ln(1 + 2\alpha)}{\alpha} \right] + \frac{\ln(1 + 2\alpha)}{2\alpha} - \frac{1 + 3\alpha}{(1 + 2\alpha)^2} \right\}$$ (III.55)

C. Pair production

Pair production is a process in which a photon gives up all its energy, E_γ, and creates an electron-positron pair (e^-,e^+). The conservation of energy requires that $E > 2m_ec^2 = 1.02$ MeV for the process to occur. Pair creation occurs predominantly in the electric field of the atomic nucleus (it is kinematically impossible for an isolated photon to create a e^--e^+ pair). Pair production is similar to the photoelectric effect: a photon lifts up an electron from a continuum of negative-energy states which are all occupied. The "hole" left behind is the positron, e^+.

The cross section for pair production increases with the energy of the incident photon and it depends quadratically on Z, the atomic number of the absorbing material. The energy available is shared approximately randomly between e^+ and e^-. The angle between e^+ (or e^-) and the direction of the incident photon is small and of the order of mc^2/T_e, where T_e is the e^+ or e^- kinetic energy.

When the pair production process occurs in the electric field of the atomic nucleus, the nucleus itself gains only little energy. However, when the process occurs in the field of an atomic electron (the threshold for this is $4m_ec^2 = 2.044$ MeV) the electron acquires enough energy and momentum to be separated from the atom; effectively, the absorption of the photon results in 3 electrons, and the process is called *triplet production*.

The positron, e^+, may - and usually does - undergo the opposite reaction:

$$e^+ + e^- \rightarrow 2\gamma. \tag{III.56}$$

This process where the positron combines with an electron of the absorber is called *positron annihilation*; its cross section increases inversely proportional to the positron energy, attaining a maximum when e^+ is at rest. The two photons of reaction, Eq(III.56), are then emitted back to back with an energy of 0.511 MeV each. Again, because of kinematical constraints, a single photon may be emitted only if the electron is tightly bound to the atomic nucleus (this single-photon process is more rare than two-γ annihilation). A selection of practical formulae for pair production cross sections may be found in the book of B. Rossi (1952).

III.4.2 Neutron-interaction Cross Sections

Neutrons undergo various types of interactions with matter. In describing this, it is here convenient to classify the neutrons according to their energy as follows:

a) *Thermal* neutrons have energies of the order of 0.025 eV. At these energies the neutron velocity is comparable with molecular velocities at room temperature (hence the name "thermal").

b) *Intermediate* neutrons with energies between thermal and 10 keV.

c) *Fast* neutrons have energies in the range of 10 keV to 20 MeV.

d) Neutrons with energy larger than about 20 MeV are termed *relativistic*.

The dominant interaction for thermal neutrons are:

$$(n, \gamma) \; radiative \; neutron \; capture \tag{III.57}$$

$$(n, p), (n, d), (n, t), (n, \alpha) \tag{III.58}$$

$$(n, \; fission \; products). \tag{III.59}$$

Reaction (III.59) occurs with heavier nuclei (e.g. U or Pu). Capture reaction with the emission of a charged particle may occur on light nuclei as well.

For intermediate neutrons the main reaction is elastic scattering:

$$(n, n) \tag{III.60}$$

Characteristic for fast neutrons is elastic and inelastic scattering:

$$(n, n') \tag{III.61}$$

where the target is left in an excited state; de-excitation occurs via photon emission. As the incident neutron energy approaches the energy of one of the excited levels of the nucleus the cross section for the reaction, Eq(III.61), increases substantially, a phenomenon known as *resonance* scattering.

At higher neutron energies (fast and relativistic neutrons) other inelastic channels open; they are collectively known as nuclear reactions. Examples are:

$$(n, n'p), (n, n'\alpha), (n, 2n) \tag{III.62}$$

An inelastic reaction such as

$$n + {}^{12}C \rightarrow n' + 3\alpha \tag{III.63}$$

leads to target fragmentation.

There exist extensive compilations of neutron cross sections and kerma factors (ICRU, 1977). An example of such data is shown in Fig.III.7. The kinematics of elastic neutron scattering has been discussed in the Example 1 of Sect. III.3.

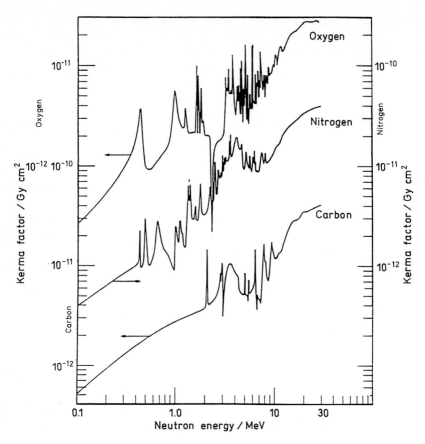

Fig.III.7 Kerma factors for carbon, nitrogen and oxygen as a function of neutron energy. Note the separate ordinate scale for each element (after Caswell et al, 1977 as quoted in ICRU, 1977).

III.4.3 Charged Particles as Sources of other Charged Particles

Electrons, protons, heavier ions or mesons may be sources of charged particles either directly (ionization, nuclear reactions) or indirectly (e.g. bremsstrahlung). A comprehensive theoretical or experimental description of these processes does not exist. In Sect. III.5 we introduce a formal framework for describing interaction probabilities for Coulomb-interaction mediated processes; as already indicated, this is the main mechanism by which energy is imparted to the medium by ionizing radiation. Here we review several exact results concerning the interaction of charged particles with *free* (rather than atomic bound) electrons, and also data on nuclear reactions. Under the conditions where binding energies may be neglected (e.g. at large incident energies or large momentum transfer) these formulae are useful for microdosimetric calculations.

Let $\hbar\omega$ represent the energy transfer by a charged particle of energy E, charge z and velocity β (in units of c) to electrons - assumed free - in a medium of atomic number Z. We denote by $d\sigma(E,\hbar\omega)\ /\ d\hbar\omega$ the single-differential cross section for this process *per atom*. The expressions below are reproduced from B. Rossi (1952).

For spin 0 projectiles:

$$\frac{d\sigma(E,\hbar\omega)}{d(\hbar\omega)} = \frac{d\sigma_R(E,\hbar\omega)}{d(\hbar\omega)}\left[1 - \beta^2\frac{\omega}{\omega_{max}}\right], \qquad \text{(III.64)}$$

where

$$\frac{d\sigma_R(E,\hbar\omega)}{d(\hbar\omega)} = \frac{2\pi r_e^2}{\beta^2}\ m_ec^2\ \frac{1}{(\hbar\omega)^2} \qquad \text{(III.65)}$$

is the Rutherford cross section and $\hbar\omega_{max}$ is the maximum energy transfer:

$$(\hbar\omega)_{max} \cong 2m_ec^2\frac{\beta^2}{1-\beta^2}. \qquad \text{(III.66)}$$

For a projectile with spin 1/2 and mass m:

$$\frac{d\sigma(E,\hbar\omega)}{d(\hbar\omega)} = \frac{d\sigma_R(E,\hbar\omega)}{d(\hbar\omega)}\left[1 - \beta^2\frac{\omega}{\omega_{max}} + \frac{1}{2}\left(\frac{\hbar\omega}{E+mc^2}\right)^2\right] \qquad \text{(III.67)}$$

The collision of an electron with (free) atomic electrons is given by the *Møller* equation (Møller, 1932):

$$\frac{d\sigma(E,\hbar\omega)}{d(\hbar\omega)} = \frac{d\sigma_R(E,\hbar\omega)}{d(\hbar\omega)}\frac{1}{(E-\hbar\omega)^2}\left[1 - \frac{\hbar\omega}{E} + \left(\frac{\hbar\omega}{E}\right)^2\right]^2 \qquad \text{(III.68)}$$

This equation is derived under the assumption that the projectile and target electrons are indistinguishable. By convention, the energy $\hbar\omega$ is restricted to (0,E/2), and the more energetic electron emerges from the collision with energy E-$\hbar\omega$. Eq(III.68) is invariant to exchanging $\hbar\omega$ and E-$\hbar\omega$. The scattering angles are given by (NCRP, 1991):

$$\sin^2\theta_1 = \frac{2\hbar\omega}{E}\left[\frac{E}{m_ec^2}\left(1-\frac{\hbar\omega}{E}\right)+2\right], \qquad \text{(III.69)}$$

$$\sin^2\theta_2 = 2\left(1-\frac{\hbar\omega}{E}\right)\left[\frac{E}{m_ec^2}\left(\frac{\hbar\omega}{E}\right)+2\right], \qquad \text{(III.70)}$$

where θ_1 corresponds to the more energetic of the two electrons.

The collision of positron projectiles is described by the *Bhabha* formula:

$$\frac{d\sigma(E, \hbar\omega)}{d(\hbar\omega)} = \frac{d\sigma_R}{d(\hbar\omega)}\left[1 - \frac{\hbar\omega}{E} + \left(\frac{\hbar\omega}{E}\right)^2\right] \tag{III.71}$$

A second mechanism by which charged particles lose energy and generate other particles is *bremsstrahlung*, which consists of emission of photons when the trajectory of the particle is deflected in the electric field of a nucleus. For a particle of mass m the bremsstrahlung probability is proportional to Z^2/m^2 and therefore, at the energies typically considered in microdosimetric applications, only electrons have significant radiation loss. Detailed formulae for this process have been given by Bethe and Heitler (1934). Tabulations of double-differential cross sections for bremsstrahlung radiation can be found in Kissel et al (1983). At large electron energies radiation energy loss (bremsstrahlung) becomes more important than collision energy loss. For electrons incident on water this happens at energies larger than about 80 MeV.

Sufficiently energetic heavy ions generate additional charged particles in nuclear reactions. The minimum energy for these processes is determined by the Coulomb threshold between projectile and target. A distinction is made between projectile fragmentation - which results in particles with approximately the same velocity as the initial particle - and spallation products which result when the target nuclei break up in fragments.

Negative pions, π^-, with energies below 100 MeV have been used as sources of charged particles as a result of the fact that - at the end of their range - they form mesonic atoms and, with high probability, are absorbed by nuclei. Their rest mass (140 MeV) is converted in the fragmentation of the nucleus into a "star" of protons, alpha particles, neutrons and other nuclear fragments. Per π^- capture in ^{12}C one obtains (Büche and Przybilla, 1981) approximately 2.5 neutrons (75%), 0.5 p, d, and α each (5-10%) and 0.2 t, Li, Be and B (0.1-2%). The numbers in parentheses indicate the percentage of the kinetic energy of 140 MeV.

III.5 Microscopic Description of the Electromagnetic Interaction of Charged Particles with Matter

Energy deposition in matter by charged particles occurs predominantly via Coulomb-field interactions. Although the formal theoretical description of this kind of interaction is well developed (see the Appendix to this chapter) numerical evaluations of cross sections remain very difficult, particularly in condensed media where one deals, simultaneously, with many-body systems containing a large number of targets. This is in direct contrast with the gas phase where isolated molecules are the targets and wave functions and eigenvalues for molecules may be obtained quite accurately with standard *self-consistent field* (SCF) methods, for instance the *Hartree-Fock* (HF) technique supplemented with corrections for the correlation energy. A similar imbalance between gas and condensed phase exists in terms of available experimental data. Our treatment of the interaction of charged particles with water vapor (see Sect. V.2.3) is almost entirely based on available experimental data. For solid state targets data are sparse and the treatment of charged particle interactions must be based to a great extent on calculations.

There are other, more fundamental phase-specific differences in the interaction of charged particles. In condensed matter charged particles may (and often do) induce *collective* (coherent) excitations, such as *plasmons* or *excitons*. These quanta are *delocalized* over macroscopic spatial regions and their decay into more conventional single-particle excitations (e.g. ionizations) need not coincide with the location of the energy-loss event. The microdosimetric consequences of this state of affairs are very important, but only limited theoretical or experimental progress has been made in this direction.

III.5.1 Theoretical Outline

It can be shown (see Appendix) that the doubly-differential cross section per atomic electron, $d^2\sigma / d(\hbar\omega)d(\hbar q)$, for a particle of mass m, velocity v and charge z, to undergo energy loss $\hbar\omega$ and momentum transfer $\hbar q$, may be related to the dielectric response function of the medium, $\varepsilon(q,\omega)$, as follows:

$$\frac{d^2\sigma}{d\omega\, dq} = \frac{2z^2 e^2}{\pi N\hbar v^2}\frac{1}{q}\,\mathrm{Im}\left[-\frac{1}{\varepsilon(q,\omega)}\right]. \qquad (\text{III.72})$$

The quantity $\mathrm{Im}[-1/\varepsilon]$ is termed the *energy loss function* of the medium. The expression, Eq(III.72), contains the simplification that momentum transfer is isotropically distributed. As written, Eq(III.72) uses the cgse system of units, that is the permitivity of the vacuum, ε_0, is taken as unity.

An equivalent form of Eq(III.72), differential in the solid angle, $d\Omega$, of the scattered particle (rather than q) is:

$$\frac{d^2\sigma}{d(\hbar\omega)d\Omega} = \frac{1}{(e\pi a_0)^2 N}\frac{p_f}{p_i}\frac{1}{(\hbar q)^2}\,\mathrm{Im}\left[-\frac{1}{\varepsilon(q,\omega)}\right]. \qquad (\text{III.73})$$

Here p_i and p_f are the initial and final momenta (i.e. $\hbar q = p_i - p_f$), a_0 is the Bohr radius:

$$a_0 = \frac{\hbar^2}{m_e e^2},$$

(III.74)

and N is the total electron density in the target. With Λ the mean-free path, one has:

$$\frac{d^2(\Lambda^{-1})}{d(\hbar\omega)d(\hbar q)} = \frac{d^2(N\sigma)}{d(\hbar\omega)d(\hbar q)}.$$

(III.75)

The energy-loss function is related to the, perhaps more familiar, differential *generalized oscillator strength* (GOS):

$$\frac{df(\hbar\omega, \hbar q)}{d(\hbar\omega)} = \frac{2\hbar\omega}{\pi(\hbar\omega_p)^2} \, \text{Im}\left[-\frac{1}{\varepsilon(q, \omega)}\right],$$

(III.76)

where $\hbar\omega_p$, the *plasma frequency*, is given by

$$\hbar\omega_p = \sqrt{\frac{4\pi Ne^2}{m}}.$$

(III.77)

$\hbar\omega_p$ depends only on the electron density, N, in the medium.

The dielectric response function, $\varepsilon(q,\omega)$, and in particular the energy loss function, are the fundamental quantities which determine for *all* phases the response of the system to the charged particle[10].

III.5.2 Experimental Data on the Energy Loss Function

Data on $d^2\sigma / d(\hbar\omega)d(\hbar q)$, and therefore $\text{Im}[-1/\varepsilon]$, are generally obtained from electron scattering experiments or from optical data. Both methods have shortcomings. Electron data are typically measured in transmission or reflection experiments. In the former, one needs very thin targets (2-200 nm), comparable with the mean-free path of the electrons. To the extent that multiple scattering occurs, unfolding algorithms are needed. On the other hand, the finite thickness needed for the target material to be self supporting limits the minimum projectile energy that can be used (for electrons, about 10 keV). To minimize the target thickness traversed by the projectile data must be taken at forward angles, that is

---

[10] The real part, ε_1, and the imaginary part, ε_2, of $\varepsilon(q,\omega)$ are not independent; they satisfy the Kramers-Kronig dispersion relations and in principle one needs only one of the two quantities to obtain the entire function, $\varepsilon(q,\omega)$.

at very small momentum transfer. Reflection experiments are, in turn, significantly affected by impurities on the surface of sample.

The optical determination of $\varepsilon(q,\omega)$ takes advantage of the relationship between the complex refractive index, $n_o(\omega) = n(\omega) + ik(\omega)$, and the dielectric function, ε:

$$n_o^2 = \varepsilon = \varepsilon_1 + i\varepsilon_2. \tag{III.78}$$

The momentum transferred in optical spectroscopy is very small, so effectively measurements of $n(\omega)$ and $k(\omega)$ as a function of photon energy, $\hbar\omega$, provide information [via Eq(III.78) and Kramers-Kronig analysis] only on $\varepsilon(q=0,\omega)$; this is termed the optical limit $(q \to 0)$ of $\varepsilon(q,\omega)$. Optical data are often obtained over a limited range of photon energies.

In Fig.III.8 we show illustrative data on ε and $Im[-1/\varepsilon]$ for water. The peak in $Im[-1/\varepsilon]$ at about 8 eV corresponds to an exciton line. The large structure (peak) which appears at about 20-30 eV in many of the measured spectra has been attributed to plasmon excitation. The validity of this interpretation remains however under dispute because the plasmon signature $(\varepsilon_1=0)$ is often not even approximately fulfilled.

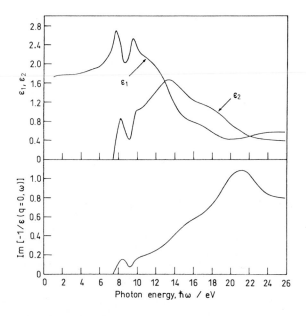

Fig.III.8 The real part, ε_1, and the imaginary part, ε_2, of the dielectric response function of water as a function of incident photon energy. The data are obtained at the optical limit $(q=0)$. Also shown is the energy loss function, $Im[-1/\varepsilon]$. (after Heller et al, 1974).

III.6   The Interaction of Charged Particles with Bulk Matter

In the previous Sect. we have discussed interaction processes in thin targets - thin referring to target dimensions small relative to the mean free path for the interactions considered. In most practical applications charged particles traverse substantially larger amounts of matter, and the overall energy deposition and/or angular deflection result from the combination of a large number of elementary interaction events. These average quantities, and their statistical fluctuations, make the subject of this section.

III.6.1 The Stopping Power of the Medium

The *stopping power* of a medium for a charged particle has been defined in Sect. III.2: it is the average energy loss per unit path length as the particle traverses the medium. A distinction is made between the *collision* stopping power, S_{coll}, which represents the result of ionizations and excitations, and the *radiative* stopping power, which is the result of photon emission associated with bremsstrahlung.

The collision stopping power for a charged particle of energy T, velocity v and charge z is defined as follows:

$$S_{coll}(T) = \int_0^\infty \hbar\omega \frac{d\Lambda^{-1}}{d(\hbar\omega)} d(\hbar\omega) = \frac{2z^2 e^2}{\pi \hbar v^2} \int d\hbar\omega(\hbar\omega) \int \frac{1}{\hbar q} \operatorname{Im}\left[-\frac{1}{\varepsilon(q,\omega)}\right] d(\hbar q).$$

$$(III.79)$$

The notations of Eqs(III.72,75) are used here. The salient features of S_{coll} appear if Eq(III.79) is evaluated under certain simplifying assumptions: At sufficiently large energies, T, one may approximate the energy-loss function with its forward scattering ($q=q_{min}$) value:

$$\operatorname{Im}\left[-\frac{1}{\varepsilon(q,\omega)}\right] \cong \operatorname{Im}\left[-\frac{1}{\varepsilon(q_{min},\omega)}\right], \qquad (III.80)$$

which thus becomes a function of ω only. From the kinematics of the collision:

$$\hbar q_{max,min} = \sqrt{2T} \pm \sqrt{2(T - \hbar\omega)}. \qquad (III.81)$$

When $T \gg \hbar\omega$

$$\hbar q_{min} \cong \hbar\omega / \sqrt{2T},$$

$$\hbar q_{max} \cong 2\sqrt{2T}. \qquad (III.82)$$

Eq(III.79) becomes:

$$S_{coll}(T) \cong \frac{2z^2 e^2}{\pi v^2} \int (\hbar \omega)(\hbar \omega) \, \text{Im}\left[-\frac{1}{\varepsilon(q_{min}, \omega)}\right] \log \frac{4T}{\hbar \omega}. \qquad (III.83)$$

This expression may be further simplified by using a sum rule, Eq(III.175, Appendix III.7), and by defining the *mean excitation energy* as follows:

$$\log I = \int d(\hbar\omega) \frac{df(q_{min}, \omega)}{d\omega} \log(\hbar\omega) = \frac{1}{2\pi^2 N} \int d(\hbar\omega)(\hbar\omega) \, \text{Im}\left[-\frac{1}{\varepsilon(q_{min}, \omega)}\right] \log(\hbar\omega)$$

$$(III.84)$$

One obtains

$$S_{coll}(T) \cong \frac{4\pi z^2 e^2 N}{v^2} \log \frac{4T}{I}. \qquad (III.85)$$

Within this approximation $S_{coll}(T)$ is proportional to the ratio z^2/v^2 characterizing the projectile, does not depend on the projectile mass, and is proportional to the atomic number of the material, Z, since

$$N = \rho \frac{N_A}{M_A} Z. \qquad (III.86)$$

(N_A=Avogadro's number, M_A=atomic mass of the target atom).

From the theory of Bethe (1930,1932,1933) the following expressions apply for heavy particles with charge ze:

$$\frac{1}{\rho} S_{coll}(T) = \frac{4\pi N_A r_e^2 mc^2}{\beta^2} z^2 \frac{Z}{M_A}\left[\log \frac{2mc^2 \beta^2 T}{I^2(1-\beta^2)} - \frac{c}{Z} - \frac{\delta}{2} + zL_1 + z^2 L_2\right]$$

$$(III.87)$$

For electrons:

$$\frac{1}{\rho} S_{coll}(T) = \frac{2\pi N_A r_e^2 mc^2}{\beta^2} \frac{Z}{M_A}[\log \frac{mc^2 \beta^2 T}{2I^2(1-\beta^2)}$$

$$-(2(1-\beta^2) - 1 + \beta^2)\log 2 + (1-\beta^2) \qquad (III.88)$$

$$+ \frac{1}{v}(1-(1-\beta^2))^2]$$

δ, c/Z, zL_1 and $z^2 L_2$ represent corrections to the Bethe's equations.

Eqs(III.87,88) are basically valid whenever the projectile's velocity is large compared with the velocity of atomic electrons.

The mean excitation energy, I, as a function of Z is shown in Fig.III.9. The radiative stopping power, $S_{rad}(T)$, becomes important only at high energies; for instance, for electrons incident on water $S_{rad} = S_{coll}$ at about $T = 75$ MeV.

At lower electron energies (e.g. < 10 keV), and the corresponding energies for heavier charged particles, the Bethe's formulae are not applicable and calculations need to be made based on evaluations (generally empirical) of the dielectric response function, as indicated in Eq(III.79). Slow[11], heavy charged particles also undergo charge transfer reactions, that is - as they move through matter - they will

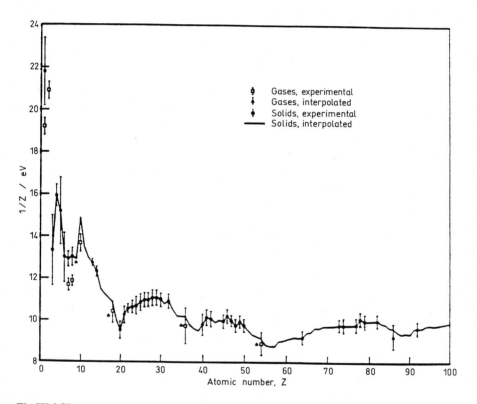

Fig.III.9 The mean excitation energy, I, as a function of the atomic number Z. (reproduced from ICRU, 1984).

---

[11] As already mentioned, "slow" means that the incident particle is moving with a velocity lower or comparable to the velocity of atomic electrons. The average value of atomic electron velocity is given by $Z^{1/3}c/137$.

pick up and then lose atomic electrons. The effective charge of these particles, z^*, is less than z and the following empirical formula has been suggested (Barkas, 1963):

$$z^* = z\left[1 - e^{-125\beta z^{-2/3}}\right].$$ (III.89)

Figure III.10 shows a plot of the mass stopping power of water for protons as a function of proton kinetic energy. There are three distinct regions in this graph: to the right of the maximum value, $S(T)$ decreases according to the Bethe's formula, that is as $1/v^2$. $S(T)$ reaches a minimum at about $\beta=0.95$. At higher energies (the third region) S increases with T because of the contribution of S_{rad}. The maximum value in S occurs at energies where the maximum energy transfer, T_{max}, is about equal to $2I$. To the left of the maximum S decreases with decreasing projectile energy: the decrease in effective charge, Eq(III.89), is partly responsible for this shape.

The stopping power for compounds may be obtained, approximately, by using the Bragg's additivity rule: for a compound, the mass collision stopping power is the sum of the mass collision powers of the atomic constituents weighted by the fractional contribution by weight of each constituent. The approximate nature of this rule stems from the fact that the chemical binding energy of the compound is not being taken into account.

The computer code SPAR (Amstrong and Chandler, 1973) may be used to calculate stopping powers for a variety of materials and incident particles.

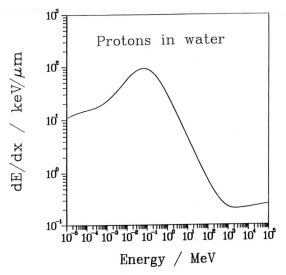

Fig.III.10 Calculated stopping power, dE/dx, of water for protons. The results have been obtained with the code SPAR (Armstrong and Chandler, 1973).

III.6.2 Statistical Fluctuations of the Energy Lost by Charged Particles

The energy lost by a charged particle in traversing a slab of material of thickness x is a stochastic quantity: it depends on two random factors, the *number* of single-collision (single ionizations or excitations), and the *energy* deposited in each collision. The former is described by a Poisson distribution; while the second is given by the single-collision spectrum, $d\sigma(\hbar\omega, T)/d(\hbar\omega)$, of the particle. The combination of these two factors is termed a *Poisson-compound process*, and it has been already discussed in Chapter II in relation to multi-event microdosimetric distributions. We shall recapitulate here this formalism as applied to fluctuations in the energy loss by charged particles. We follow the derivation of Kellerer (1985).

Let W represent the average energy lost by the particle as it traverses the slab of material. The number of collisions along the track, v, is distributed according to the Poisson's formula since single collisions are statistically independent events. The probability distribution of v is:

$$p(v) = e^{-W/w_{IF}}\left(\frac{W}{w_{IF}}\right)^{v}\frac{1}{v!},$$

(III.90)

where w_{IF} is the mean (frequency) energy loss per primary collision:

$$w_{IF} = \int w\frac{d\sigma}{dw}dw \; / \; \int\frac{d\sigma}{dw}dw$$

(III.91)

and W/w_{IF} is the mean number of primary collisions per particle traversal. In complete analogy with Eq(II.7) we can express the probability, $f(w)dw$, of energy loss in $[w,w+dw]$ as

$$f(w) = \sum_{v=0}^{\infty} e^{-W/w_{1F}}\left(\frac{W}{w_{1F}}\right)^{v}\frac{1}{v!}f_{v}(w)$$

(III.92)

where [see Eqs(II.6)] $f_v(w)$ is the probability of energy loss w in exactly v primary collisions. Explicitly:

$$f_0(w) = \delta(w)$$

$$f_1(w) = \frac{d\sigma(w)}{dw} \; / \; \int\frac{d\sigma(w)}{dw}dw,$$

(III.93)

$$f_v(w) = \int_0^w f_{v-1}(w - w')f_1(w')dw'.$$

The variance of the straggling distribution, f(w), is given by [see Eq(II.24)]

$$\sigma^2_w = w_{1D}W \qquad \text{(III.94)}$$

where

$$w_{1D} = \int \frac{w^2 f_1(w)}{w_{1F}} dw \qquad \text{(III.95)}$$

Eq(III.92) provides a complete description of the straggling distribution.

An important simplification occurs when the average number of primary collisions satisfies: $W/w_{1F} \gg 1$. Then Eq(III.92) may be well approximated with a Gaussian (normal) distribution function:

$$f(w; W) \cong \frac{1}{\sqrt{2\pi w_{1D}W}} e^{-\frac{(w-W)^2}{2w_{1D}W}} \qquad \text{(III.96)}$$

The notation indicates explicitly the W-dependency of f(w). Note also that W is proportional with the thickness x of the slab.

Analytic expressions for $f(\omega)$ have been obtained under various simplifying assumptions (Landau, 1944; Vavilov, 1957). For instance, taking $f(w)=w^{-2}$ one obtains Vavilov's approximation

$$w_{1D} = \frac{w_{max}}{2 \log(w_{max} / I)}, \qquad \text{(III.97)}$$

where the integral of Eq(III.95) was performed between $w_{min}=I/w_{max}$ and $w_{max}=(m_e/m)T$, [see Eq(III.32)]. Vavilov's approximation is valid only for charged particles colliding with free electrons; it is not applicable when the energy loss, w, is comparable with the binding energy.

Finally, it may be noted that because of the asymmetric shape of f(w) the most probable energy loss [i.e. the peak of f(w)] is smaller than the average energy loss, W.

III.6.3 Range and Range Straggling

Particles of a given energy, T, for which the energy straggling function is relatively narrow will have traversed approximately the same distance by the time they are brought to rest. One may then introduce the notion of particle *range*.

Formally (see Eq.III.13):

$$R(T) = \int_0^T dx = \int_0^T \frac{dT'}{S(T')}. \qquad \text{(III.98)}$$

R is the *average* range of the particles and S(T) is the linear stopping power of the medium. There is no exact analytic form for R(T); however, the following approximate expression may be used (ICRU, 1978):

$$R(T) = \frac{k}{z^2} \frac{T^{1.73}}{m^{0.73}} \qquad \text{(III.99)}$$

for $10^{-3} < T/mc^2 < 0.8$. z and m are, respectively, the charge and mass of the particle and k is a material dependent constant. A range- energy curve for protons in water is shown in Fig.III.11.

Corresponding to energy straggling one has the phenomenon of *range straggling*. The range fluctuation is approximately Gaussian with standard deviation, σ_R, given by (Rossi, 1952)

$$\sigma_R = \sqrt{\frac{m_e}{m}} \, g\!\left(\frac{T}{mc^2}\right), \qquad \text{(III.100)}$$

where g is a slowly varying function of T.

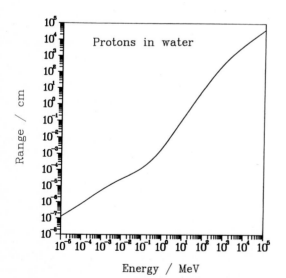

Fig.III.11 The range of protons in water. The results have been obtained with the code SPAR (Armstrong and Chandler, 1973).

III.7 Appendix

Formal Treatment of the Interaction of Charged Particles with Matter

A detailed description of the interactions of charged particles with matter is beyond the charter of this book. Nor would such a description be particularly needed, as most microdosimetrists are basically terminal-end users of expressions (e.g. cross sections) describing the interaction processes. It appears relevant nevertheless to introduce the basic concepts utilized in the description of charged-particle scattering in a way which is sufficiently general to be applicable to both condensed *and* gas-phase systems, thus providing a starting point for the study of more detailed treatments, should the reader be so inclined.

The main results of Sect. III.7.1 are Eqs(III.140,147,148) that give quantum mechanical expressions for the cross sections differential in energy and momentum transfer. More details about these derivations may be found in the books by Wu (1962) and Goldberger and Watson (1964) on which the material in Sect. III.7.1 is based. Sect. III.7.2 contains derivations of an exact expression for the dielectric response function, Eq(III.167), and an explicit relation between this quantity and the cross sections described in the previous Sect. [Eq(III.171)]. A complete treatment of these topics has been given by Pines and Nozieres (1966). Finally, Sect. III.7.3 gives some examples of practical ways to perform calculations of the dielectric response functions, and therefore obtain cross sections necessary for describing particle transport in matter. Natural units ($\hbar = 1$, $m_e = 1$, $c = 1$) are used whenever convenient.

III.7.1 Scattering Formalism

Some simplifying assumptions are used in obtaining the results below. These approximations are common to most calculations and have here the additional advantage of making most of the concepts introduced more transparent.

Consider a system, S, consisting of a projectile and a target. Let H_p and H_t, respectively, be the Hamiltonian operators for the projectile and target when isolated from each other. Thus:

$$H_p = \frac{p^2}{2m},$$

(III.101)

where m is the mass of the projectile and p the momentum operator (operators are represented by *bold*-face letters). Let further V represent the interaction potential between projectile and target. The Hamiltonian for the system is:

$$H = H_p + H_t + V.$$

(III.102)

Eigenvectors and eigenvalues for these operators are designated as follows:

$$H_t|E_i \ge E_i|E_i >,$$ (III.103)

$$H_p|q \ge \frac{q^2}{2m}|q >,$$ (III.104)

$$H|\Psi \ge E|\Psi >.$$ (III.105)

Note the following: a) the projectile is represented - when far away from the target - by a plane wave (free particle with definite energy and momentum). This means that the projectile is not localized in either space or time and therefore a *time-independent* treatment of the problem is anticipated. This is, of course, a simplification: In real situations the projectile has a certain degree of localization (at the expense of corresponding uncertainties in energy and momentum); using however a wave packet, rather than Eq(III.104), makes the problem significantly more complicated.

b) H_t and H_p act in *different* ket spaces, and H acts in a space which is the direct product of spaces span by $|E_i>$ and $|q>$.

Consider now the following expansion:

$$|\Psi >= \sum_n |\Phi_n > |E_n >.$$ (III.106)

The coefficients of this expansion, $|\Phi_n>$, belong to the projectile space. From Eqs(III.105) and (III.106):

$$\sum_n (H_t + H_p)|\Phi > |E_n > + V|\Psi >= E \sum_n |\Phi_n > |E_n >.$$ (III.107)

Multiply both terms by $<E_f|$ and use Eq(III.103) and the fact that $|E_i>$ are orthonormal:

$$(E - E_f - H_p)|\Phi_f >= \sum_n < E_f|V|E_n > |\Phi_n >.$$ (III.108)

At large distances (away from the scattering region) $V=0$ and $|\Phi_f>\to|q>$ (a free particle), with $q^2/2m = E-E_f$. A formal solution of Eq(III.108) is:

$$|\Phi_f \ge |q > + \frac{1}{E - E_f - H_p} \sum_n < E_f|V|E_n > |\Phi_n >.$$ (III.109)

It is not an actual solution since $|\Phi>$ appears on both sides of Eq(III.109); it also requires a definition of the operator $(E-E_f-H_p)^{-1}$. This is examined in the following.

To simplify the notation define a wave number, k, by setting

$$E - E_f = \frac{k^2}{2m}. \tag{III.110}$$

We are interested then in the operator $(k^2 - p^2)^{-1}$ where p is the momentum operator, $p|p> = p|p>$. Let $|\Phi>$ be a state in $|q>$-space (i.e. momentum space). Then using the relation

$$\sum_p |p><p| = 1 \tag{III.111}$$

one obtains:

$$\frac{1}{k^2 - p^2}|\Phi> = \frac{1}{k^2 - p^2}\sum_{p'}|p'><p'|\Phi> =$$

$$= \sum_{p'}\frac{1}{k^2 - p'^2}<p'|\Phi>|p'> \tag{III.112}$$

In the position representation

$$<r|\Phi> = \Phi(r), \tag{III.113}$$

$$<r|p> = \frac{1}{(2\pi)^{3/2}}e^{ipr}, \tag{III.114}$$

$$<p|\Phi> = \int \frac{1}{(2\pi)^{3/2}}e^{-ipr}\Phi(r)dr, \tag{III.115}$$

one obtains from Eqs(III.112-115):

$$\frac{1}{k^2 - p^2}\Phi(r) = \int dr\left[\int \frac{1}{(2\pi)^3}\frac{1}{k^2 - p^2}e^{ipr-ipr'}dp\right]\Phi(r') \tag{III.116}$$

The quantity in square brackets is the *Green's function*:

$$G(r) = \frac{1}{(2\pi)^3}\int \frac{e^{ipr}}{k^2 - p^2}dp. \tag{III.117}$$

It is clear that for real values of $\mathbf{p}$ this integral is not defined (it is divergent at p=k). It can be shown - by performing the angular integration and using the theorem of residues - that this difficulty is removed if the poles of p are made slightly complex.

$$G(r) = \frac{1}{(2\pi)^3} \int \frac{e^{ipr}}{(k + i\eta)^2 - p^2} \, d\mathbf{p}, \tag{III.118}$$

meant in the limit of $\eta \to 0$. The result is:

$$G(r) = -\frac{1}{4\pi} \frac{e^{ikr}}{r}, \tag{III.119}$$

as is necessary in order to have the scattered particle represented by a spherical wave.

The correct solution of Eq(III.108) is then:

$$|\Phi_f > = |q > + \frac{1}{E - E_f - H_p + i\eta} \sum_n < E_f|V|E_n > |\Phi_n >, \tag{III.120}$$

rather than Eq(III.109). In the position representation this becomes [see Eqs(III.116,119)]:

$$\Phi^+_f(r) = \frac{1}{(2\pi)^{3/2}} e^{iqr} - 2m \int dr' \frac{e^{ik|r-r'|}}{4\pi|r - r'|}$$
$$< E_f|V|\Psi^+(r') >, \tag{III.121}$$

where we have used the notation:

$$|\Psi^+(r) > = \sum_n \Phi^+_n(r)| E_n >. \tag{III.122}$$

The integration domain in the integral of Eq(III.121) is determined by the range of V. For r>>r' and with the notations $\hat{r}=r/r$, $\hat{k}=k\hat{r}$ we have

$$|r - r'| \to r - \hat{r}r', \tag{III.123}$$

$$k|r - r'| \to kr - \hat{k}r', \tag{III.124}$$

$$\Phi^+_f(r) \to \frac{1}{(2\pi)^{3/2}} e^{iqr} - \frac{2m}{4\pi} \frac{e^{ikr}}{r} \int dr' e^{-\hat{k}r'}$$
$$< E_f|V|\Psi^+(r') >, \tag{III.125}$$

$$\Phi^+{}_f(r) \rightarrow \frac{1}{(2\pi)^{3/2}} e^{iqr} - \frac{e^{ikr}}{r}\left[\frac{2m}{4\pi} < k, E_f|V|\Psi^+>\right]. \qquad \text{(III.126)}$$

The ket $|k, E_f>$ represents a state made of a scattered particle of momentum k and the target left with energy E_f. The square bracket contains the *amplitude* of the scattered wave, $f(q,k)$.

The result, Eq(III.126), is exact. To make it more explicit, however, successive iterations (called the 1-st, 2-nd, ... Born approximations) are made. Thus the first term of the Born series (this is most commonly used) is obtained by making the substitution:

$$|\Psi^+(r)> \cong \frac{1}{(2\pi)^{3/2}} e^{iqr}|E_i>. \qquad \text{(III.127)}$$

This means [see Eq(III.122)] approximating $\phi^+(r)$ by the first term of Eq(III.126) and assuming the initial target state $|E_i>$. The result is

$$f(q, k) = \frac{2m}{4\pi} < k, E_f|V|q, E_i >. \qquad \text{(III.128)}$$

Finally, we calculate the double differential cross section with respect to solid angle and energy, $d\sigma/d\Omega d\omega$. By definition:

$$\frac{d\sigma}{d\Omega} d\Omega = \frac{\textit{fluence trough area } d\sigma = r^2 d\Omega}{\textit{incident fluence}} \qquad \text{(III.129)}$$

For a particle (mass m) with wave function Ψ the flux, j, is obtained from the continuity equation:

$$\frac{\partial \rho}{\partial t} + \nabla j = 0, \qquad \text{(III.130)}$$

where $\rho(r,t)=|\Psi|^2$, and Ψ satisfy:

$$-\frac{\hbar^2}{2m} \nabla^2\Psi + V\Psi = i\hbar \frac{\partial \Psi}{\partial t}. \qquad \text{(III.131)}$$

Thus

$$j = \frac{i\hbar}{2m}(\Psi^*\nabla\Psi - \Psi\nabla\Psi^*), \qquad \text{(III.132)}$$

as can be easily verified by inserting the expression, Eq(III.132), back in Eq(III.130). The numerator in Eq(III.129) is calculated using the second term on the r.h.s of Eq(III.126), while the first term is used to calculate the denominator. The result is:

$$\frac{d\sigma(i \to f)}{d\Omega} = \frac{k}{q}\left|\frac{2m}{4\pi}\right| < k, E_f|V|q, E_i >\Big|^2 .$$

(III.133)

Conservation of energy allows only particles with energy $\omega = E_f\text{-}E_i$ to emerge from the collision process. Then

$$\frac{d^2\sigma(i \to f)}{d\Omega\, d\omega} = (2\pi)^6 \frac{k}{q}\left|\frac{2m}{4\pi}\right| < k, E_f|V|q, E_i >\Big|^2 \delta(\omega + E_i - E_f).$$

(III.134)

The expression, Eq(III.134), can be further simplified. Assume that the interaction potential, **V**, can be put in the form:

$$V = \sum_j V(r_j - r),$$

(III.135)

where **r** and r_j are, respectively, the position vectors for the projectile and particle j of the target. Then

$$< k, E_f|V|q, E_i > = \frac{1}{(2\pi)^3} \int dr e^{i(q-k)r}$$

$$< E_f|\sum_j V(r_j - r)| E_i > .$$

(III.136)

On the other hand,

$$\sum_j \int dr e^{ipr} V(r - r_j) = \sum_j \int dr e^{ip(r+r_j)} V(r)$$

$$= V(p)\sum_j e^{ipr_j} = V(p)\rho^+(p).$$

(III.137)

Here $p = q\text{-}k$ = momentum transfer and V(p) and ρ(p) are the Fourier transforms of V and $\rho(r_j)$, respectively. Indeed,

$$\rho(p) = \int \rho(r)e^{-ipr} dr = \int \left[\sum_j \delta(r - r_j)\right] e^{-ipr}$$

$$= -\sum_j e^{-ipr_j}$$

(III.138)

Thus

$$< k, E_f |V| q, E_i > = \frac{1}{(2\pi)^3} V(p) < E_f | \rho^+(p)| E_i > \qquad \text{(III.139)}$$

and

$$\frac{d^2\sigma(i \rightarrow f)}{d\Omega\,d\omega} = \frac{k}{q} |V(p)|^2 |< E_f | \rho^+(p)| E_i >|^2 \, \delta(\omega + E_i - E_f) \qquad \text{(III.140)}$$

The quantity

$$F_{i\rightarrow f}(p) = \frac{1}{N} < E_f | \rho^+(p)| E_i >, \qquad \text{(III.141)}$$

where N is the total number of target particles, is known as the *form factor* of the system for a transition with momentum transfer **p**. Also

$$f_{i\rightarrow f}(p) = \frac{2m}{p^2} (E_f - E_i) |< E_f | \rho^+(p)| E_i >|^2 \qquad \text{(III.142)}$$

is called the *oscillator strength* for the transition i→f. It can be shown that:

$$\sum_{if} f_{i\rightarrow f} = N, \qquad \text{(III.143)}$$

a useful sum rule indicating the relative "strength" of each particular transition. To the extent that the final state of the target, f, is unknown (or just irrelevant to the problem) one can sum Eq(III.140) over all allowed final states. It is convenient to replace Ω by p as follows:

$$p^2 = |k - q|^2 = k^2 + q^2 - 2kq\cos\theta, \qquad \text{(III.144)}$$

$$dp = -\frac{kq}{p} d(\cos\theta) = -\frac{kq}{p} \frac{d\Omega}{2\pi}, \qquad \text{(III.145)}$$

$$\frac{d^2\sigma}{d\Omega\,d\omega} = \frac{d^2\sigma}{dp\,d\omega} \frac{kq}{2\pi\,p}. \qquad \text{(III.146)}$$

The expression, Eq(III.140), becomes

$$\frac{d^2\sigma(p,\omega)}{dp\,d\omega} = \frac{2\pi\,p}{q^2} |V(p)|^2 \, S(p,\omega) \qquad \text{(III.147)}$$

where, by definition, the *dynamic form factor* is

$$S(p, \omega) = \sum_{if} |< E_f| \rho^+(p)| E_i >|^2 \, \delta(\omega + E_i - E_f). \qquad \text{(III.148)}$$

Note that Eq(III.147) is the product of two factors, one containing information on the projectile-target interaction [$V(p)$], the other describing the target [$S(p, \omega)$].

The expression just obtained, Eq(III.147), contains - in principle - all that is needed to calculate elastic and inelastic scattering probabilities. It is customary, however, to replace the dynamic form factor by a quantity which is calculationally more convenient (and perhaps easier to interpret), namely, the dielectric response function of the system, $\varepsilon(q,\omega)$.

III.7.2 The Dielectric Response Function

The *dielectric response function*, $\varepsilon(k,\omega)$ is defined in terms of the electric field **E** and the dielectric displacement **D** induced by a charged projectile of charge density $\rho_p(r,t)$:

$$E(k, \omega) = D(k, \omega) \, / \, \varepsilon(k, \omega). \qquad \text{(III.149)}$$

This relation is the Fourier transform of

$$D(r, t) = \int dr'dt' \varepsilon(r' - r', t - t') E(r', t') \qquad \text{(III.150)}$$

which shows that **D** is related to **E** through a linear but *non-local* relation (i.e. the dielectric displacement at **r**,t relates to values of the electric field at previous times and different locations).

From Maxwell's equations

$$\nabla D(r, t) = 4\pi \, \rho_p(r, t) \qquad \text{(III.151)}$$

$$\nabla E(r, t) = 4\pi [\rho_p(r, t) + < \rho_t(r, t) >]. \qquad \text{(III.152)}$$

Here $<\Delta\rho_t>$ is the average charge fluctuation induced in the target system by the projectile. From these equations one obtains:

$$\frac{1}{\varepsilon(k, \omega)} = 1 + \frac{< \Delta\rho_t(k, \omega) >}{\rho_p(k, \omega)}. \qquad \text{(III.153)}$$

In order to evaluate ε we need the ratio $<\Delta\rho_t>/\rho_p$. This is calculated as follows:

Assume the effect of the projectile on the target sufficiently week to be able to treat it as a (linear) perturbation. "Linear" response means that we can treat each Fourier component of ρ_p separately. Without any loss of generality let then

$$\rho_p(r, t) = \rho_p(k)e^{i(kr-\omega t)+i\delta} + c.c. \qquad (III.154)$$

The positive quantity δ (understood to be taken in the limit $\delta \to 0$) is introduced as an artificial device to insure that the projectile and target do not interact at $t \to \infty$. It also allows to treat ρ_p as a *time-dependent* perturbation. The Coulomb interaction between ρ_p and ρ_t is:

$$V(r_t, t) = \int \frac{\rho_p(r, t)\rho_t(r_2, r_t)}{|r_1 - r_2|} dr_1 dr_2$$

$$= \rho_p(k) \int \frac{\rho_t(r_2, r_t)\exp[i(kr_1 - \omega t) + i\delta t]}{|r_1 - r_2|} dr_1 dr_2 + c.c. \qquad (III.155)$$

$$V(r, t) = \frac{4\pi}{k^2} \rho_p(k)e^{\delta t - i\omega t} \int \rho_t(r_2, r_t)e^{ikr} 2 dr_2 + c.c.$$

$$= \frac{4\pi}{k^2} \rho_p(k)\rho_t^+(k, r_t)e^{-i\omega t + \delta t} + c.c. \qquad (III.156)$$

The notation $\rho_t(r_2, r_t)$ refers to the charge density produced by target particles (represented by the generalized coordinate r_t) at position r_2.

The problem at hand is to find $<\Delta\rho(k,\omega)>$ for a system (the target) under the influence of a time-dependent perturbation [the projectile, via the interaction potential $V(r_t, t)$ of Eq(III.156)]. We need to find therefore the wave function of the target. We shall work in the position representation. Let H_t be the (stationary) Hamiltonian of the target *in the absence* of any perturbation:

$$H_t\Phi_j(r) = E_j\Phi_j(r) \qquad (III.157)$$

When $V(r_t, t)$ is turned on the Schrödinger equation becomes:

$$ih\frac{\partial}{\partial t}\Psi(r, t) = [H_t + V(r, t)]\Psi(r, t). \qquad (III.158)$$

(the subscript t is no longer used). Consider next the Green function $G(r,r',t)$ defined as a solution of

$$\left[\frac{h}{i}\frac{\partial}{\partial t} + H_t\right]G^+(r', r, t) = h\delta(r' - r)\delta(t). \qquad (III.159)$$

The relation between this expression and Eq(III.158) becomes evident by rewriting this latter as

$$\left(\frac{h}{i}\frac{\partial}{\partial t} + H_t\right)\Psi(r,t) = -V(r,t)\Psi(r,t). \tag{III.160}$$

Once we have a solution of Eq(III.159) the wave function Ψ can be calculated from

$$\Psi(r,t) = \Phi(r,t) + \int G^+(r',r,t-t')V(r',t')dr'dt', \tag{III.161}$$

as can be readily verified. The solution of Eq(III.159) is:

$$G^+(r',r,t) = \frac{1}{2\pi}\sum_j \int dE e^{-iEt}\frac{\Phi^*_j(r')\Phi_j(r)}{E - E_j} \tag{III.162}$$

and consequently:

$$\Psi(r,t) = \Phi(r,t) + \frac{1}{2\pi}\sum_j \int dE dr'dt' e^{-iE(t-t')}\frac{\Phi^*_j(r')V(r',t')\Psi(r',t')}{E - E_j}\Phi_j(r) \tag{III.163}$$

This integral equation can be solved by successive iterations. To first order we shall simply approximate $\Psi(r',t')$ on the r.h.s. of Eq(III.163) with its unperturbed value $\phi_0(r',t')$. We shall further assume that originally the target was in a pure state

$$\Phi_0(r,t) = \Phi_0(r)e^{-iE_0 t}. \tag{III.164}$$

Then

$$\Psi(r,t) = \Phi_0(r,t) + \frac{1}{2\pi}\sum_j \int \frac{dE}{E-E_j}e^{-iEt} < \Phi_j|V_0|\Phi_0 > \Phi_j(r,t)\int dt'\, e^{-i(E_0-E+\omega+i\delta)t'}$$

$$= \Phi_0 + \sum_j \int \frac{dE}{E-E_j}e^{-iEt} < \Phi_j|V_0|\Phi_0 > \delta(E_0 - E + \omega - i\delta)\Phi_j(r)$$

$$= \Phi_0 - \frac{4\pi}{k^2}\rho_p(k)\sum_j < \Phi_j|\rho^+_t(k)|\Phi_0 > \frac{e^{-i\omega t+\delta t}}{\omega_{j0} - \omega - i\delta}\Phi_j(r)e^{-iE_0 t} \tag{III.165}$$

Here, $V_0 = (4\pi/k^2)\rho_p(k)\rho^+_t(k)$ and $\omega_{j0} = E_j - E_0$. For clarity the second term (c.c.) in Eq(III.156) has been omitted here. Now:

$$< \Delta\rho(k, \omega) >=< \Psi| \, \rho_t(k, \omega)|\Psi >$$

$$= -\frac{4\pi}{k^2} \, P_p(k) \sum_j |< \Phi_j| \, \rho_t(k)|\Phi_0 >|^2 \left[\frac{1}{\omega_{j0} - \omega - i\delta} + \frac{1}{\omega_{j0} + \omega + i\delta} \right].$$

$$(\text{III.166})$$

and from Eq(III.153):

$$\frac{1}{\varepsilon(k, \omega)} = 1 - \frac{4\pi}{k^2} \sum_j |< \Phi_j| \, \rho_t(k)|\Phi_0 >|^2 \left[\frac{1}{\omega_{j0} - \omega - i\delta} + \frac{1}{\omega_{j0} + \omega + i\delta} \right].$$

$$(\text{III.167})$$

This is the desired expression for $\varepsilon(\mathbf{k},\omega)$ in terms of the eigenvectors of the target.

In this notation Eq(III.148) reads:

$$S(k, \omega) = \sum_j |< \Phi_j| \, \rho_t^+(k) |\Phi_0 >|^2 \, \delta(\omega - \omega_{j0}) \qquad (\text{III.168})$$

and using the identity

$$\lim_{\delta \to 0} \frac{1}{x \pm i\delta} = P\left(\frac{1}{x}\right) \mp i\pi\delta(x) \qquad (\text{III.169})$$

(P = principal value) one obtains

$$\text{Im}\left[\frac{1}{\varepsilon(k, \omega)} \right] = \frac{4\pi}{k^2} [S(k,-\omega) - S(k, \omega)] \qquad (\text{III.170})$$

Finally, for positive energy transfer ($\omega > 0$) Eq(III.147) becomes:

$$\frac{d^2\sigma(k, \omega)}{dk\,d\omega} = \frac{2}{q^2} \frac{1}{k} \text{Im}\left[-\frac{1}{\varepsilon(k, \omega)} \right]. \qquad (\text{III.171})$$

As a remainder, q here is the initial momentum of the projectile.

The expression, Eq(III.171), is the desired relation between the double differential cross section (in energy and momentum transfer) and the dielectric response function of the target. Although kinematic constraints relate $\mathbf{k}$ to ω, at large projectile energies (relative to ω) the correlation between these two quantities can be neglected.

III.7.3 Theoretical Calculations of the Energy Loss Function

The expression, Eq(III.167), is an exact expression for $\varepsilon(\mathbf{q},\omega)$ for homogeneous, isotropic systems. The calculation of $\varepsilon(\mathbf{q},\omega)$ requires, therefore, a quantum-mechanical description of the band structure of the system and its wavefunctions, a task beyond current computational capabilities for any but the simplest materials. A number of alternate approaches have been proposed; they are reviewed below. These approximate methods are all based on the premise that optical data, $\varepsilon(q=0,\omega)$, are available from experiment and therefore one needs the theoretical expression to extrapolate ε to non-zero momentum transfers.

III.7.3.1 Drude-function Expansions of $\varepsilon(q,\omega)$

The starting point for this approach is the following expression (termed the Drude formula) of $\varepsilon(q,\omega)$:

$$\varepsilon(0, \omega) = 1 + \sum_n \frac{f_n \omega_p^2}{\omega_n^2 - \omega^2 - i\gamma_n \omega} . \tag{III.172}$$

f_n is the oscillator strength of a transition with energy ω_n, and γ_n are damping parameters. A variation of this formula was used by Ritchie et al (1988) as follows:

$$\varepsilon_2(0, \omega) = \omega_p^2 \sum_n \frac{2 f_n \gamma_n \omega^3}{\left[\left(\omega_n^2 - \omega^2\right) + \gamma_n^2 \omega^2\right]^2} \tag{III.173}$$

$$\varepsilon_1(0, \omega) = 1 + \omega_p^2 \sum_n \frac{f_n\left[\omega_n^2 - \omega^2\right]\left[\left(\omega_n^2 - \omega^2\right)^2 + 3\gamma_n^2 \omega^2\right]}{\left[\left(\omega_n^2 - \omega^2\right)^2 + \gamma_n^2 \omega^2\right]^2} . \tag{III.174}$$

In Eqs(III.173,174) γ_n, ω_n and f_n are fitting parameters obtained by matching these expressions to available optical data. The fitting procedure is constrained by *sum rules*, for instance:

$$\int_0^\infty \omega \, \mathrm{Im}\left[-\frac{1}{\varepsilon(q, \omega)}\right] d\omega = \frac{\pi}{2} \omega_p^2 . \tag{III.175}$$

The parameters ω_n and γ_n are assumed to be q-dependent in a way that leads to $\omega_n(q) = q^2/2$ at large values of q (the Bethe ridge).

There are other model dielectric functions (Mahan, 1990) but one clearly puts a premium on computational simplicity, when selecting one of them. Drude expressions appear to satisfy this desideratum.

III.7.3.2 Random-phase Approximation (RPA) for $\varepsilon(q,\omega)$

Ehrenreich and Cohen (1959) have derived the following expression for $\varepsilon(q,\omega)$:

$$\varepsilon(q,\omega) = 1 + \frac{4\pi e^2}{q^2 \Omega} \sum_{knl} \frac{|F_{qnl}(k)|^2}{\omega_l(k+q) - \omega_n(k) - \omega + i\delta}, \qquad \text{(III.176)}$$

where the form factor F is given by:

$$F_{qnl}(k) = <l, k+q|e^{-qr}|n, k>. \qquad \text{(III.177)}$$

As opposed to the exact expression, Eq(III.167), here wave functions $|n,k>$ and energy bands $\omega_n(k)$ are obtained in the single-particle approximation. n and k represent, respectively, the band index and the wave vector. The summation is over occupied states, n, and unoccupied states, l.

Several levels of approximation have been used in calculating Eq(III.176): In a model developed by Ritchie and Tung (Ritchie et al, 1988) the occupied states are represented by simple atomic orbitals, while the conduction states are taken to be orthogonalized plane waves (OPW):

$$u_c(k, r) = \frac{e^{ikr}}{\sqrt{\Omega}} - \sum_k \sum_n \left[\int \frac{e^{ikr}}{\sqrt{\Omega}} u_n(q, r')dr' \right] u_n(q, r), \qquad \text{(III.178)}$$

where c=conduction, n=occupied. Typically one takes an s orbital for u_n:

$$u_n(q, r) = \sqrt{\frac{\alpha_n^3}{\pi}} e^{-\alpha_n r} \qquad \text{(III.179)}$$

and assumes that

$$\omega_n(q, r) = \beta_n q^2. \qquad \text{(III.180)}$$

The expression, Eq(III.176), can be then obtained in closed form and α_n and β_n used as adjustable parameters.

One could further improve the quality of the calculation with Eq(III.176) by deriving more realistic wavefunctions and energy bands. As an example, a calculation of $\varepsilon(\mathbf{q},\omega)$ for cubic ice (as a model for condensed water) uses a semiempirical Hamiltonian and solves the Schrödinger equation of the system (Zaider et al, 1990).

III.7.3.3 Ab initio Calculations of $\varepsilon(\mathbf{q},\omega)$

Atomic hydrogen is the only system for which $\varepsilon(\mathbf{q},\omega)$ is exactly known. For other systems one may use SCF Hartree-Fock calculations to obtain the ground state energy and good quality wave functions. However, excited (unoccupied) states energies may be in error by as much as 50% due to the neglect of electron corre-lation. A simple possibility of correcting this is to use the so-called *scissors operator,* which amounts to displacing vertically the conduction bands by the amount necessary to obtain agreement with spectroscopic data.

A truly *ab initio* approach will involve the calculation of quasi-particle excitations of the system. Although theoretical methods for doing this are well developed (Hedin and Lundqvist, 1969) their practical implementation is restricted to very simple systems.

Chapter IV
Experimental Microsimetry

IV.1 The Site Concept

The observation that ionizing radiations can differ greatly in effectiveness indicates that local concentrations of absorbed energy must be of cardinal importance. The obvious way of assigning a quantitative meaning to the term "local concentration of absorbed energy" is to define it as the energy absorbed in volumes of specified dimensions, and the dimensions of interest are those of the regions in the irradiated material where the concentration of absorbed energy determines the probability of a given effect. In preliminary treatments these regions can be regarded to be convex geometric volumes of equal size that are termed *sites*. *Regional microdosimetry* is concerned with energy deposition in sites and it is the principal object of experimental microsimetry.

Since the local concentration of absorbed energy is governed by the rate of energy loss in the tracks of charged particles, their LET has commonly been considered to be the quantity to which differences in effectiveness are to be attributed. The connection can be only qualitative because of the complex relation between the average rate of the energy expended in the track of a charged particle and the energy *absorbed* in a site that is traversed by the particle.

Even with the extreme simplification in which it is assumed that the particle trajectory is rectilinear, that the energy loss is continuous, constant, and confined to the track; and that the particle range is large compared to the dimensions of the site, the LET can only characterize the mean energy deposited in traversals because of the varying track length intercepted by the site. As already mentioned, by a theorem due to Cauchy, $\bar{l}$, the mean chord length in random[12] traversal of a body is 4 V/S where V and S are its volume and surface. In the case of a spherical site of diameter d the mean energy, $\bar{\varepsilon}$, deposited by particles of LET, L is thus given by

---

[12] In this context, which refers to isotropic irradiation and uniform fluence, the term *random* refers to the so-called μ-*randomness* (Kellerer, 1980).

$$\bar{\varepsilon} = \bar{l}L = L\frac{2d}{3} \tag{IV.1}$$

In many cases the finite range, R, of the particle must be taken into account. A theorem by Kellerer (1980) provides a formula for $\bar{s}$, the mean track length:

$$\bar{s} = \left(\frac{1}{\bar{l}} + \frac{1}{\bar{R}}\right)^{-1}. \tag{IV.2}$$

where $\bar{R}$ is the average value of R. However one needs to go beyond these simple relations. When R is comparable to d, L can not be considered to be constant and the appropriate integration may require knowledge of the change of L along the track which depends on the mass and the charge, i.e. the nature, of the particle.

Further difficulties are illustrated in Fig.IV.1 a and b which shows two-dimensional projections of the tracks of two 1-keV electrons in water vapor and makes it apparent that other characteristics complicate the situation. One of these is *track curvature* which results in a larger energy deposit as compared with a rectilinear track. This is less important for smaller values of d, but for these the *energy escaping* from a site, in the form of delta radiation[13] becomes more significant.

Finally, a process that is generally of overriding importance and that is at the core of microdosimetry is *straggling* which can result in extreme variations of energy deposition. Charged particles lose energy discontinuously and in varying amounts resulting in fluctuations that are nearly always significant and that become extreme in small sites. In sites having dimensions that are comparable to the diameter of the DNA molecule the importance of straggling and the consequent independence of energy absorption from LET is evident in the fact that in a *thousand-fold* increase of LET from its minimum value to 100 keV/μm the magnitude of energy deposition varies by a factor that is scarcely more than 2 (Rossi, 1993).

Kellerer and Chmelevsky (1975) have provided graphs presenting the range of diameters of the sites in which the mean energy deposited by various particles is within 10% of the value that may be calculated on the basis of LET. In the case of protons of energies, E_p, between 0.2 and 6 MeV, d can not differ by a factor of more than 3 from

$$d = 0.9E_p^{1.35} \tag{IV.3}$$

---

[13] The commonly employed term "delta ray escape" refers to the dominant process and should be understood to also include the usually much smaller transmission of energy by fluorescence photons.

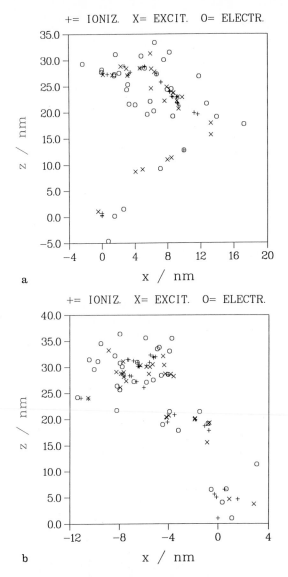

Fig.IV.1 a and b. Two dimensional view of two 1-keV electron tracks. The individual ionizations, excitations and thermalized electrons are shown with the symbols indicated.

and corresponding bands are calculated for alpha particles and heavier ions. For electrons no such bands exist at all.

Beyond these important shortcomings there is a further fundamental deficiency of LET because it can in any case only provide information on the *mean* energy absorbed. Consequently, one can infer the probability of only those effects that are proportional to absorbed energy. But many radiation effect probabilities and especially those for most biological effects are *not* proportional to absorbed energy.

In an attempt to overcome the limitation of the LET concept the quantity has been defined as L_Δ where Δ represents an energy limit beyond which delta rays are considered to be particles that furnish a separate contribution to the LET distribution in irradiated material. This excludes the initial portion of the trajectories of these particles from the energy deposited "locally". Although an appropriate choice of Δ can permit a closer approximation to the energy imparted to a site, the fundamental objection remains that this can only be a mean value that has limited utility, especially if small sites are involved.

These difficulties are obviated by the stochastic concepts of lineal energy, y, and specific energy, z, because they are defined in terms of the energy actually absorbed and therefore also are the quantities most readily measured. With some exceptions (Section IV.5) the measurement is one of the probability density distribution of lineal energy, f(y). As shown in Chapter II this is closely related to the corresponding distribution of z as produced in single events. The increment of specific energy, Δz, produced in a spherical region of diameter d by an event of lineal energy y is

$$\frac{\Delta z}{Gy} = 0.204 \frac{y}{keV\,\mu m^{-1}} \left(\frac{d}{\mu m}\right)^{-2} \tag{IV.4}$$

The definition of y is in terms of the mean site diameter and it has the same physical dimension as L. Figure IV.2 shows calculated distributions f(y) for 125 keV protons in spherical regions of various diameters. The even wider distributions for 100 keV electrons are shown in Fig.IV.3.

The definition of y is applicable to sites of all shapes but it is usually considered, and most frequently measured, in *spherical* sites. This is the only shape for which f(y) is independent of the angular distribution of the radiation. Since non-spherical sites are often randomly oriented (e.g. chromosomes in biological irradiations) f(y) should, in principle, be determined for isotropic irradiations of non-spherical sites. f(y) distributions are not the same for equal average chord lengths $\bar{l}$ in shapes that need to be specified by more than one parameter because chord length distributions differ. However, as will be shown below, the path length distribution is often a minor contributor to the shape of f(y).

While the specification of radiation quality in terms of lineal energy distributions provides a realistic statement on energy deposition, it contains a number of disadvantages. The most obvious one is that, unlike L which refers to radiation without reference to circumscribed regions in irradiated material, y must necessarily refer at least to the mean site diameter. As seen in Figs.IV.2 and IV.3,

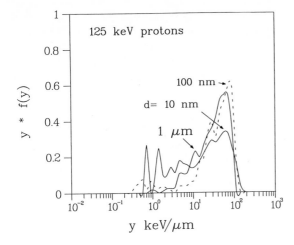

Fig.IV.2 Calculated lineal energy distributions of 125 keV protons for site diameters 10, 100 and 1000 nm.

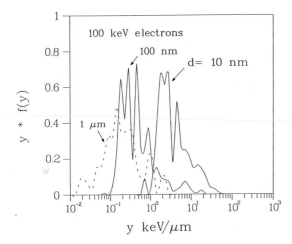

Fig.IV.3 Calculated lineal energy distributions of 100 keV electrons for site diameters 10, 100 and 1000 nm.

f(y), although reasonably constant when changes in d are moderate, differs substantially when they are large.

Another, generally more important, limitation of lineal energy (and the related specific energy) distributions is the fact that they do not provide information on the pattern of energy deposition *within* the site. Strictly speaking y is sufficient for a correlation between radiation and its effects only in two extreme cases: either the energy transfer points are *randomly* distributed in the site, or the effect is due to changes in target entities that are *randomly* distributed in the site and interact with a probability that is independent of their initial separation. The first of these conditions is never met for ionizing particles that have energies in excess of 100 eV. In VI.1.4 the second condition is assumed in an approximation.

The more advanced formulation of radiation quality developed by Kellerer is based on proximity functions considered in Chapter V. Lineal energy is conceptually, as well as historically, only a first step in the microdosimetric approach to radiological science but it has nevertheless proven to be a concept of appreciable utility.

IV.2 Fluctuations in Regional Microdosimetry

Kellerer (1970) has provided a thorough analysis of the random factors that determine the shape of the pulse height distribution observed when a proportional counter is traversed by charged particles. This analysis includes the distribution of energy absorbed in the gas, as well as the modification of this distribution by the measuring system. The analysis is based on the *relative variance*, V, of distributions, f(x), defined by

$$V = \frac{\sigma^2}{m_1^2} = \frac{m_2}{m_1^2} - 1 \qquad\qquad (IV.5)$$

where σ^2 is the *variance* (i.e. the square of the standard deviation, σ), and m_1 and m_2 are the first and second moments of f(x).

V is a simple index of the width of a distribution. It is dimensionless and has the advantage that the *total relative variance*, V_T, produced in a chain of independent processes can be obtained simply by adding the relative variances of each process involved. Another important consideration is that a simple relation involving the relative variance and the mean value of the single event distribution of specific energy is of primary importance in many radiation effects (see Chap. VI).

This section deals with the random factors that determine regional microdosimetric distributions. Their modification by measurement are considered in Section IV.3. The total variance, V_T, is due to six processes that are subject to statistical fluctuations and consequently characterized by probability density distributions.

Neglecting additional factors that are frequently unimportant in small sites (such as change of LET during traversal or finite particle range), they are:

A. The distribution of the *number of (energy deposition) events.*
B. The *LET distribution* of the particles.
C. The distribution of the *path lengths* of particles in the site.

A, B and C are collective properties of all particles. The three additional causes of fluctuations involve energy deposition by individual particles.

D. The distribution of the *number of collisions.*
E. The distribution of *energy imparted in individual collisions.*
F. The distribution of the fraction of this *energy retained in the site* (i.e. not escaping as delta radiation).

A: v, the number of events in the site conforms to a Poisson distribution

$$p(v) = e^{-n}\left(\frac{n^v}{v!}\right) \qquad\qquad (IV.6)$$

where n is the mean (expectation) value of v. The relative variance, $V_v = 1/n$. Most microdosimetric measurements determine f(y) the distribution of lineal energy. They, therefore, involve single events and $V_v = 0$. (Note that "single events" differs from an "average of one event"). A related distribution $f_1(z)$ specifies the distribution of specific energy in single events. However, f(z;D) the distribution of specific energy due to the absorbed D always involves multiple events.

B: The linear energy transfer, L, of particles is subject to a distribution that varies greatly depending on the radiation. In the case of monoenergetic charged particles, V_L, the relative LET variance, is equal to zero. In the more important cases of the LET spectra of charged particles set in motion by uncharged particles, V_L varies considerably. It is about 0.3 for Co γ radiation and 0.8 for 2 MeV neutrons with larger, but not very well known figures for higher neutron energies. For site diameters larger than 0.5 μm, it is a significant contribution to the total variance in neutron measurement but it contributes negligibly to V_T in photon measurements (see below).

C: It is generally believed, but unproven, that V_t, the relative variance of the intercepted length of the particle track in convex bodies is minimal in a spherical site. For particles completely traversing such a site the distribution of path length, x, is equal to

$$f(x) = \frac{x}{2r^2} \qquad\qquad (IV.7)$$

where r is the radius of the sphere. The first moment, i.e., the mean length of traversal is 4r/3. The second moment, which is proportional to the mean energy deposited by particles of uniform constant energy loss, is $2r^2$ and V_t = 1/8. If the length of tracks is comparable to the site diameter V_t is less and approaches zero for very short tracks.

The sphere is the only volume with isotropic response. Spherical geometry is also appropriate for homogenous sites where a sphere can represent a bounded, or an average, volume in which radiation products form or interact. In radiobiology one may wish to know the frequency of events, or the energy deposited in the nucleus of the cell which can often be approximated as an oblate spheroid. On the other hand, unless the sensitive volume of a microdosimetric counter is spherical, it is usually cylindrical.

The chord length distribution for isotropic incidence of rectilinear tracks may be of interest even when the radiation is directional. Unless cells are exposed to monodirectional radiation in the flattened states characteristic of tissue culture (where V_t is small) they can commonly be assumed to be randomly oriented and their average response is then that of single cells in an isotropic radiation field. Fig.IV.4 shows V_t in isotropic fields for spheroids of various elongation (Kellerer, 1985).

Since microdosimetry is nearly always performed with single detectors, the isotropic response of a cylindrical counter in a non-isotropic field would require a complex rotation at constant fluence rate. This is impractical and employment of such counters is usually based on the assumption that directional characteristics of the field can be neglected. This is always the case when $V_t \ll V_T$, a condition commonly encountered. Fig.IV.5 shows V_t in isotropic fields for cylinders of various elongation. For radiation parallel

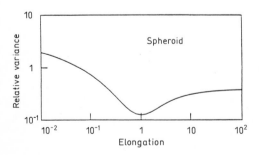

Fig.IV.4 Relative variance of chord length in oblate and prolate spheroids. The elongation is the length of the axis of circular symmetry relative to the circular diameter (after Kellerer, 1985).

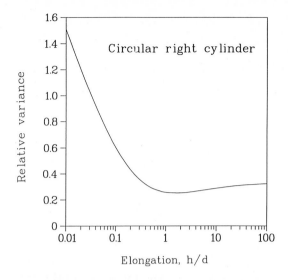

Fig.IV.5 Relative variance of the distribution of chord length in circular right cylinders as a function of elongation (the ratio of the height of the cylinder to its diameter).

or orthogonal to the cylinder axis V_t is independent of elongation and respectively equal to 0 and 0.0808.

The remaining processes constitute three aspects of *straggling*.

D: The number of collisions is Poisson - distributed about the mean value $c = L / \bar{\varepsilon}$ where $\bar{\varepsilon}$ is the average value of energy deposits (see also section III.6.2). V_c, the relative variance is therefore $1/c$.

The two remaining sources of fluctuations involve complex mechanisms which, even for given LET, depend on the velocity, mass and charge of particles.

E: There is uncertain information on the shape, $f(\varepsilon)$, the distribution of energy imparted, ε, at low values. It is well established that

$$f(\varepsilon) \approx \frac{1}{\varepsilon^2} \ for \ \frac{I^2}{\varepsilon_{max}} < \varepsilon < \varepsilon_{max} \tag{IV.8}$$

where ε_{max} is the maximum energy and I the mean ionization potential of the medium. ε_{max} depends on the mass and velocity of the charged particle and can be evaluated rather simply. However, in most collisions ε is less than the lower limit above. I is about 70 eV for water and this is much higher than the mean energy deposited at relevant transfer points (see Section IV.3.3).

F: Because of the tortuous trajectory of low energy electrons delta ray escape is another difficult problem. It involves not only $f(\varepsilon)$ but also the geometry of the site.

The total relative variance of a microdosimetric spectrum is

$$V_T = V_v + V_L + V_t + V_L V_t + V_c + p V_\varepsilon \qquad (IV.9)$$

where V_ε is the relative variance of ε and $p(>1)$ a factor accounting for δ ray escape. Because of inadequate information on these quantities V_T must be determined by measurement or calculation for each combination of radiation and site diameter.

The relative importance of random processes is illustrated in two examples. Fig.IV.6 shows the lineal energy spectrum obtained when a wall-less counter is exposed to a uniform fluence of alpha particles (Glass and Braby, 1969). It is dominated by the triangular path length distribution in the sphere. There is however even in this case a marked modification by straggling. The small pulses are due to delta rays injected by alpha particles that pass just outside of the sphere.

The other extreme is illustrated in Fig.IV.7 which is based on a study by Kellerer (1968). It shows that in calculations of the microdosimetry of ^{60}Co γ radiation

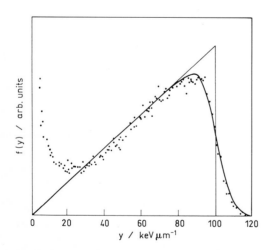

Fig.IV.6 Theoretical chord length distribution in a sphere (solid line) and experimental linear energy spectrum for 5.3-MeV alpha particles in a 1 μm diameter site (after Glass and Braby, 1969).

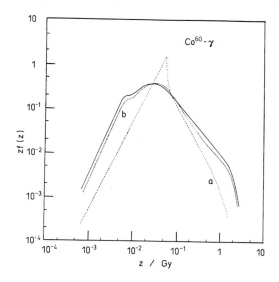

Fig.IV.7 Specific energy distribution from ^{60}Co gamma radiation in 1 μm sites. Solid curve: actual calculated distribution. Dashed curve(a): spectrum calculated on the basis of track-length and LET distributions but omission of straggling. Dotted curve(b): spectrum calculated on the basis of straggling with the omission of track length and LET distributions (after Kellerer, 1968).

neglect of the LET distribution and of the path length distribution causes only a minor change of the spectrum while neglect of straggling causes a dramatic change (note that the scales are logarithmic).

IV.3 Measurements in Regional Microdosimetry

IV.3.1 General Considerations

Like all measurements of physical quantities, experimental microdosimetry is subject to inaccuracies. It is, therefore, necessary to consider their importance as well as their causes.

As in the, generally simpler, measurements of absorbed dose, accuracy can be considered as an end in itself and there can then be no limits to demands for its improvement. On the other hand, practical applications can warrant less stringent requirements.

Radiation protection is one of the principal fields in which microdosimetric measurements are needed. It has been recommended that the quality factor, Q,

(Sect. VI.3) be based on lineal energy (ICRU, 1986) and if Q is based on LET, the most general experimental approach consists in deriving D(L), the distribution of absorbed dose in LET from its distribution in lineal energy, D(y). Even the deficiencies of this method (see Sect. IV.6) together with those in the measurement of D(y) are small compared to the uncertainties in determining the effective dose equivalent (see Sect. VI.3) in persons exposed to ionizing radiation.

Another field in which microdosimetry is of major utility is radiobiology and especially theoretical radiation biophysics. Here the accuracy requirements are intermediate because physical data on the energy distribution in irradiated tissues are related to experimental radiobiological data which have varying accuracy which is often no better than $\pm 10\%$.

In radiotherapy absorbed doses of electromagnetic radiation need to be accurate to a few percent. The same kind of accuracy is, of course, also required in high-LET (especially neutron) irradiations and it can usually be attained - especially if the absorbed dose is determined with tissue equivalent ionization chambers. There are, however, major difficulties with regard to the quality of the radiation. One of these involves the invariable admixture of gamma radiation with its lower biological effectiveness. Variations in radiation quality are usually unimportant in x- or gamma radiation therapy. However, they can be significant in neutron therapy where a related question is the distribution of neutron energies incident on the patient. This is of importance when clinical results obtained at different institutions are compared because the spectra produced by their respective accelerators are usually different. Carefully performed microdosimetry can obviate the need to know the neutron energy distribution and may provide the physical information with the needed accuracy but there is at this time no sufficiently firm information on the relation between lineal energy spectra and RBE. Consequently, microdosimetric data can be employed only in semi-quantitative considerations. However, this may change in the future.

Other areas in which accurate microdosimetric information may be needed in the future are radiation chemistry and interaction of radiation with solid state systems. Despite its evident importance, microdosimetry has thus far only rarely been utilized in the theoretical treatment of these subjects.

A major consideration in an assessment of the adequacy of measurement accuracy is the fact that (as shown in Section IV.4) microdosimetric distributions tend to be broad and deficiencies of the measurement system frequently introduce only minor distortions.

Virtually all microdosimetric measurements employ proportional counters (IV.3.2). These instruments are subject to inherent limitations as well as technical imperfections. In many cases the relative variance introduced can be calculated or measured which permits appropriate corrections. The fluctuations include that of the number of ions produced by events depositing equal energy (IV.3.3).

Since the quantity measured, i.e. y, is defined in terms of absorbed energy this variability may be considered to be a source of error which is due to varying apportionment of energy between excitation and ionization in the counter gas. However, this occurs to a similar extent in irradiated tissues. The effect thus results in fluctuations beyond those encompassed by microdosimetric quantities and is considered further in IV.3.3. Its occurrence in the counter may be regarded to be a cause of at least approximate realism rather than a deficiency. However it rarely modifies the measured spectrum significantly.

Other fluctuations are due to the varying number of electrons produced in the multiplication process (IV.3.4) and the noise in the electronic amplification system (IV.3.9). The importance of the distortions involved depends somewhat on circumstances. In general, they are more important for small events in small sites. In many practical applications the relative variance involved can be considered to be small compared to that of the true microdosimetric spectrum.

Various other methods of experimental microdosimetry have been investigated. In one of these an irradiated gas volume is *separate* from the region in which any amplification takes place. Electrons (or ions) are to be extracted, accelerated and detected by solid state detectors or multipliers (Pszona, 1976). Since the amplifying device can generally not be shielded from penetrating radiation it must be relatively small or operate in a vacuum. This poses severe technical difficulties and practical instruments utilizing this approach have not been described.

Another alternative method of experimental microdosimetry employs low pressure cloud chambers in which the 3 dimensional pattern of droplets caused by individual ionizations is resolved and stereoscopically photographed (Stonell et al., 1993). This technique is rather difficult and therefore restricted to research objectives but it is of considerable interest in that role because it provides information on the relative location of relevant transfer points for gas ionization (see Sect. V.4.3.3).

Other radiation detectors such as scintillation crystals, various solid state devices, photographic emulsions, are not capable of detecting single ionizations.

IV.3.2 The Proportional Counter

Proportional counters are the principal instruments of microdosimetry. Simulation of microscopic regions in solids by geometrically similar gas volumes of equal effective dimensions (i.e. equal products of diameter and density) avoids the problem of determining energy absorption in micrometer sized volumes.

In the gas volume the same processes occur at much larger relative distances. This magnification that is typically equal to a factor of 10^5 also results in a vast increase in the particles that are intercepted. In the usual situation where the range of charged particles in the gas is long compared with the cavity diameter the event frequency increases as the square of the magnification and the counter therefore registers the events taking place simultaneously in some 10^{10} regions in the high density material.

A further fundamental advantage is *sensitivity*. Proportional counters embody a relatively simple method of detecting a few or even single ionizations. The RMS noise in the best electronic detectors (preamplifiers) corresponds to the collection of pulses of about 100 electron charges but single electrons liberated by ionization of the gas in the counter can commonly be multiplied in an avalanche by a factor of more than 10^3. Under proper operating conditions this occurs only in the immediate vicinity of a thin anode wire with the result that the signal is essentially independent of the location of the initial ionization in the counter. It is relatively uniform internal amplification that makes the proportional counter very useful.

The detailed mechanism of proportional counter operation is complex and only few and somewhat limited investigations have been published on theoretical design criteria for microdosimetry. A pragmatic reason for the failure to pursue this subject in detail has been the fact that even early models designed with an empirical approach appeared to work satisfactorily in simulation of sites down to diameters near 0.25 μm and there appeared to be fundamental reasons why adequate accuracy could not be attained for smaller sites.

The principles involved in measurements with proportional counters as well as their design and operation are considered in the following sections IV.3.3-IV.3.11.

IV.3.3 Energy loss versus Ionization

Microdosimetric quantities and, except for exposure, all dosimetric quantities are based on energy. On the other hand the most sensitive measurements in dosimetry and especially in microdosimetry are those of ionization. The material quantity W which is termed the *mean energy expended in a gas per ion pair formed* is the quotient of the kinetic energy of a charged particle and the mean number of ion pairs that it produces in a gas. About one half of the energy of charged particles is expended in excitations of gas molecules with the other half expended in ionizations. Dosimetry or microdosimetry based on gas ionization (i.e. relevant energy transfers) is therefore a measurement of absorbed energy in integral multiples of W. In conventional dosimetry this is a negligible source of error. In absorbed dose rate measurements the ionization current is hardly ever less than one fA (femtoampere) and when the absorbed dose is measured by integration of the ion current the charge collected is usually more than 100 fC (femtocoulomb). This corresponds respectively to some $6x10^3$ charges per second and a total of $6x10^5$ charges. Other fluctuations or different sources of error are almost always more significant.

Among these is the value of W and its dependence on the type and initial energy of the ionizing particles. A review of this topic is made in an ICRU Report (1979). Srdoc et al. (1993) have provided a more recent compilation. For electrons W is nearly constant down to about 100 eV. In most cases they expend only a small fraction of their initial energy below this limit. In air W of electrons of energy above 1 keV is known to be within about 1/2 percent of 33.85 eV. In other gases employed in dosimetry, such as tissue equivalent mixtures the uncertainty is of the order of a few percent. The difference between the W values for protons and for alpha particles, and the uncertainty of either of them is also of the order of a few percent. At energies of a few MeV, where most of the data have been obtained, W of these particles is about 10% larger than for electrons. The energy dependence is more important than for electrons; there is an appreciable increase of W below energies of the order of 10 keV or more. For heavier ions W increases substantially below 1 MeV. At higher energies the W values are comparable to those of the lighter ions. Figure IV.8 shows W values for electrons and protons as well as carbon, nitrogen and oxygen ions in methane based TE gas.

W is equal to the quotient of the initial energy of a particle and the number of ion pairs produced. It therefore includes the, generally predominant, fraction of ionizations produced by delta radiation and largely represents their energy expenditure per ionization. Differences in W for various particles can therefore be attributed primarily to differences in mean delta-ray energy.

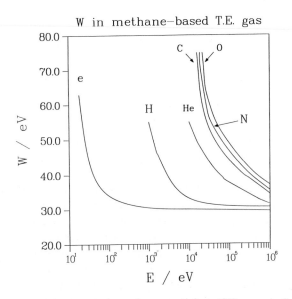

Fig.IV.8 W in methane tissue-equivalent (TE) gas as a function of the energy of electrons, protons, He, C, O and N ions.

The findings can be summarized by stating that W varies little (if at all) when the velocity of any particle is much greater than that of orbital atomic electrons and it increases when these velocities become comparable. Adopting a W value intermediate between those of fast (~1 MeV) electrons and slow (~100 KeV) protons the energy imparted to a gas can be determined in most practical situations involving protons and neutrons with an uncertainty of the order of 5% from a knowledge of the number of ion pairs produced. Knowledge of the radiations involved can of course reduce this uncertainty and more appropriate values can be chosen if e.g. only photons or neutrons are measured. In the latter case even more accurate information might be utilized. Thus calculations (Goodman and Coyne, 1980) show that W in a tissue equivalent gas mixture decreases (erratically) from 32.8 to 31 eV as the energy of neutrons increases from 0.1 to 20 MeV.

Microdosimetric spectra often convey information on the types of charged particles causing ionizations and choices of the values of W can then also be made that are more specific. There is another uncertainty in the ratio of the energies deposited per unit mass in a gas and in a surrounding solid wall. Since virtually all microdosimetric detectors are largely homogenous (i.e. gas and wall have, at least approximately, the same atomic composition) the relative stopping power in the Bragg-Gray relation is near 1. However this ratio can be somewhat larger because of the *density* (or *polarization*) *effect* which depends on the velocity of the charged particles. It is often no more than a few percent but it approaches 15% for 20 MeV electrons (Kim, 1973). At 42 MeV substantial corrections are required (Lindborg, 1974).

The most important errors in the determination of absorbed energy by ionization measurement must occur when the number of ionizations is small. This situation is peculiar to microdosimetry and unavoidable when individual energy increments in small regions of matter are assessed. In tissues or tissue equivalent materials which are of principal interest the discrepancies become substantial at dimensions of the order of 1μm for fast electrons or hard gamma radiation. In traversing a counter simulating a 1μm site, a 1 MeV electron can lose several eV in each of several collisions without producing an ionization, and hence no signal in a proportional counter. In dimensions of the order of nanometers such effects are important for all ionizing radiations.

Although the energy requirement for ionization limits the precision with which a proportional counter can be employed to determine the *energy absorbed* in a gas, the *number of ions* may well be a better index of radiation efficacy, particularly in radiobiology.

As outlined in Sect. I.2 it is apparent that only a portion of the energy imparted by ionizing radiation is instrumental in its effects, leading to the concept of *significant energy deposits*, in excess of a minimum value, ω, required for the creation of *relevant transfer points*. In the case of gas ionization, the ions are produced at relevant transfer points. Within accuracy that is sufficient for these semi-quantitative considerations W in tissue equivalent gases may be taken to be equal to 30 eV. About half of this is expended in excitations.

There may be different energy requirements for various biological effects although the primary importance and uniform construction of the DNA molecule may make a single value more probable. It is known that ultraviolet light of wave length in excess of 200 nm is biologically much less effective than ionizing radiation. Irradiation of protein films with slow electrons (Hutchinson, 1954) and of proteins and nucleic acids with photons in the vacuum ultraviolet region (Setlow, 1960) showed a sharp increase of effectiveness beginning near 10 eV which the authors tentatively identified with the onset of "ionization".

Thus there are reasons for believing that the energy expended in single changes is similar for radiobiology and tissue equivalent gases. The variance in the number of relevant transfer points is in either case smaller than the variance of absorbed energy. This is most obvious in the case where an energetic particle traversing a small volume of tissue equivalent gas produces excitations but not even one ionization. The same type of traversal would probably also not produce a relevant transfer point in the corresponding unit density volume of tissue.

The number of ions produced by an event in the counter gas is subject to fluctuations that contribute to the relative variance of the observed spectrum. Fano (1947) has shown that the relative variance is given by

$$\left(\frac{\sigma_n}{n}\right)^2 = \frac{F}{n} \qquad\qquad (IV.10)$$

where n is the average number of ions and F, known as the *Fano factor*, usually is less than one. The distribution is thus narrower than a Poisson distribution. A typical value of F is 0.3. Actual distributions of the number of ions produced in gases irradiated with soft monoenergetic x rays are shown in Fig.IV.9 (Obelic, 1985).

IV.3.4 Gas Multiplication

Microdosimeters usually have spherical cavities and the anode (commonly a thin stainless steel wire) is often surrounded by a coaxial helix in order to produce a more uniform electric field in the vicinity of the anode[14]. A sufficient

---

[14] A simpler (but somewhat less effective) method employs shaping electrodes (Benjamin et al, 1968).

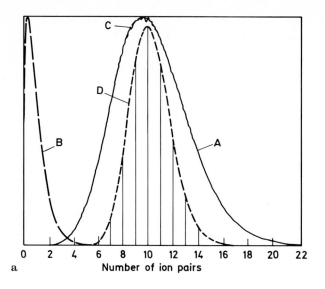

a

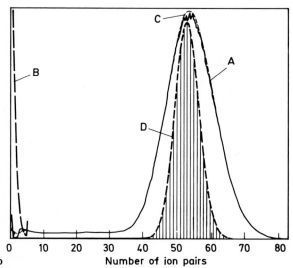

b

Fig.IV.9 (a and b) Deconvolution of the measured lineal energy spectra produced by C-K$_\alpha$ **a** and Al-K$_\alpha$ **b** x radiation. A: measured spectra, B: single electron spectra, C: calculated shape of the experimental spectrum based on the distribution of the relative number of ion pairs shown as D (after Obelic, 1985).

approximation can be made by assuming that in proper operation the multiplication occurs within the small volume encompassed by the helix in a radial electric field that is axially symmetric. Virtually all counters that are not spherical have cylindrical geometry.

An elementary analysis of proportional counter operations is generally based on the assumption of exponential growth of the electron avalanche as expressed by

$$G = \frac{N}{N_0} = e^{\alpha d} \tag{IV.11}$$

where the gain, G, is the ratio of N, the number of electrons, resulting from multiplication and N_0, the initial number, accelerated over a distance, d. α is termed the (first) Townsend coefficient. In a first approximation

$$\frac{\alpha}{p} = A e^{-Bp/X} \tag{IV.12}$$

where p is the gas pressure, X is the electric field and A and B are constants. In the (non-SI) system of units widely adopted d is measured in cm, p in torr (mm of Hg) and X in Vcm^{-1}. For methane based TE gas Campion (1972) determined A to be about 10 cm^{-1} $torr^{-1}$ and B about 210 Vcm^{-1} $torr^{-1}$.

A simple physical picture underlying this treatment is that A is the reciprocal of λ, the mean free path for ionizing collisions (and thus the mean number of such collisions per cm) at unit pressure; and B is the quotient of the mean ionization potential, I, of the gas and λ. When (X/p) λ, the mean energy acquired between collisions is large compared to I, each collision results in ionization and α/p saturates. There are a number of reasons why Eq.(IV.11) is not entirely accurate (von Engel, 1965). In particular α actually peaks at a value that is usually near 10 ion pairs cm^{-1} $torr^{-1}$ in the neighborhood of 1 kV cm^{-1} $torr^{-1}$. This behavior was observed in a proportional counter employed in microdosimetry (Schmitz and Booz, 1989) and it is one of the major reasons why the simple treatment given here must be considered an approximation.

At somewhat higher field strength *electric breakdown* (a continuous discharge) occurs in the gas. This is largely caused by the liberation of additional electrons at the cathode and in the gas by secondary processes (such as the production of photoelectrons by UV emitted in fluorescence following ionization). Quantitative, or even detailed qualitative, information on these effects is not available for TE solids and gases.

The standard analysis for cylindrical counters is thus of limited validity but it provides some guidance for counter design. The electric field at radius r in cylindrical geometry is given by

$$X = \frac{V}{r \ln \frac{c}{a}} \tag{IV.13}$$

where c and a are the radii of the cathode and anode respectively and V is the applied potential. With Eq(IV.12) an integration results in

$$\ln G \approx \int_a^c \alpha \, dr = \frac{AV}{B \ln \frac{c}{a}} \left[e^{-ap \frac{B \ln \frac{c}{a}}{V}} - e^{-cp \frac{B \ln \frac{c}{a}}{V}} \right]. \tag{IV.14}$$

In microdosimetry the energy deposition in a tissue volume of diameter d is determined by measurement in a geometrically similar TE gas volume of diameter $d(\rho_T/\rho_G)$ where ρ_T is the density of tissue and ρ_G the density of gas which is proportional to p. Within the range of its applicability Eq IV.14 indicates that if, in simulation of a given tissue volume, the relative counter dimensions are kept constant a given gain requires the same voltage regardless of the physical size of the counter because c/a, ap and cp remain the same.

Furthermore, X/p, the electric field per unit pressure, also remains the same since a change by a factor of K of the distances c,a and r in Eq IV.13 requires a change of pressure by a factor of 1/K in an equivalent geometry. Since breakdown depends at least approximately on X/p these considerations suggest that the physical size of the counter has little influence on the limits of its operation if a given tissue volume is simulated and the relative dimensions of the electrodes are the same. This is of practical importance because dimensional changes are needed to obtain adequate but not excessive counting rates for a wide range of intensities and types of radiation. As indicated above, in most applications the counting rate depends on the square of the counter diameter. As will be shown below, spherical counters ranging in diameter from 1/4 to 8 inches have been utilized accommodating a thousand fold range of relative counting rates. Employment of largely equivalent, cylindrical models down to 1/16 inch extends the range by a further factor of 16. However the restricted number of available wire diameters necessitates only approximate scaling.

Equation IV.14 can also serve to demonstrate at least the general nature of the difficulties encountered in attempts to simulate increasingly smaller regions of unit density tissue.

In proper operation of a proportional counter, gas multiplication occurs only in the immediate vicinity of the anode so that electrons liberated anywhere in nearly all the counter volume are equally multiplied. However, as the pressure is reduced the multiplication region extends to increasing radial distances if the gain is kept the same. Furthermore, it is generally necessary to increase the gain because at lower pressures charged particles lose less energy in the counter. These difficulties are especially significant in microdosimetry where often only a few ionizations occur in tenuous gas volumes.

The gain when an electron originates at a radial distance r from the center of the cylindrical counter can be derived from Eq IV.14. A numerical example applies to a cathode radius, c, of 1.5 mm and an anode radius, a, of 0.00125 cm (1 mil wire). These are typical dimensions of the helix and the center wire in a spherical counter of 1.25 cm radius in which a pressure of 28 torr of methane based TE gas simulates a 1 μm diameter site. The field between the helix and the counter shell varies axially but in the vicinity of the helix it is, in this semi-quantitative example, assumed to be a continuation of the radial field inside the helix. Equation IV.14 then represents the gain when an electron originates at a distance r from the center wire if the second term in the parenthesis is changed to exp(-rp B ln(r/a)V⁻¹). At 28 torr of methane T.E. gas a gain of 1000 requires a potential of about 730 V between helix and center wire and the gain remains constant outside of the helix. Thus, except for a small cylindrical volume around the center wire there is virtually equal gain for electrons produced anywhere in the counter (Fig.IV.10). If a 0.75 μm site is simulated at a pressure of 21 torr (and nearly the same potential) this is essentially also the case. However the further, seemingly modest, reduction of the site diameter to 0.5 μm (at a pressure of 14 torr) not only requires an appreciable increase of potential at lower pressure (a trend that must ultimately lead to breakdown) but the gain of 1000 at the helix radius of 0.15 cm results in a maximum gain nearly twice as large occurring at a radius that is about twice as large. In order to achieve essentially constant gain

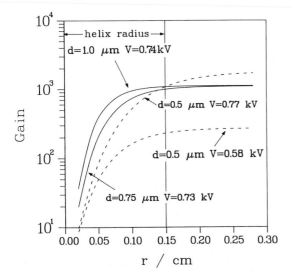

Fig.IV.10 Radial dependence of gain according to Eq(IV.14) in a proportional counter under various operating conditions. For further details see text.

beyond the helix the potential must be reduced to 580 V with a maximum gain of 200. Furthermore, an event of given lineal energy usually produces only half as many electrons as in a 1 μm site. Hence, if the multiplication is to be confined to the helix volume and an equal signal is to be available for pulse height analysis, the electronic gain must be increased by a factor of 10 with attendant instability.

Because of the scaling characteristics discussed above, operation at higher pressures and the attendant requirement of smaller cathode diameter results in equal limitations. Thus the employment of smaller counters may be expected to result in lower counting rates but not in better information on microdosimetry in smaller sites.

These considerations support the general experimental observation that adequate performance of proportional counters of conventional size (>1 cm) is limited to simulated diameters of the order of a few tenths of a micrometer. However, Kliauga (1990) reports to have obtained acceptable spectra in simulated site diameters as low as 5 nm in a very small (0.5 mm) cylindrical counter and this cannot be dismissed on theoretical grounds because the elementary treatment of gas multiplication can lead to appreciable errors as pointed out by various investigators and especially Segur et al.(1990).

One complication is lack of equilibrium between the electrons and the local electric field. This is especially pronounced in strong field gradients that exist near anodes because the electrons can then not sufficiently accelerate between collisions to reach the energy which they would acquire in a uniform field. The deficiency is accentuated by space charge that distorts the electric field and is due to differential mobility of electrons and the much slower ions. It is evident that the magnitude of the latter effect must depend on the LET of the charged particle and be more pronounced at high counting rates. It must also depend on the orientation of the particle track and be less when it is axial (parallel to the anode) rather than radial (perpendicular to it).

A further major complication is due to the fact that the motion of the electrons is not strictly radial. At low pressures they may circle the wire and undergo one or more further collisions before being absorbed in the wire. Unlike the previous effects this causes a higher multiplication.

Other factors influencing counter operation are the non-negligible production of various ionized species (e.g. CH_3^+, CH^+, H^+), energy losses by excitation, and photoionization.

Such phenomena are taken into account in the more refined theoretical treatment by Segur et al. (1990) who show much better agreement between theory and experiment. However these authors have not provided detailed guidance on counter design.

Additional reasons why Eq IV.14 should be considered to be no more than a rough guide include the facts that α depends on X/p (Engel, 1965) and that there can be interference between the avalanches triggered by single electrons. This should be less important when the trajectory of the charged particle causing the event is parallel rather than orthogonal to the anode wire. There is little information on the details of this process but Srdoc (1970) has stated that as a rule of thumb the total number of electrons reaching the anode should not exceed 10^8 for strict proportionality between the number of ion pairs and the charge collected at the anode.

In view of the complexities involved it is not surprising that experimental investigations indicate an appreciable divergence from the predictions of the usual theoretical treatment. An example provided by Srdoc (1970) is shown in Fig.IV.11 as curves of multiplication vs. voltage in propane based tissue equivalent gas. These data were obtained with a 2.5 cm diameter counter and show the need for a sizable increase in the applied potential when equal gain is to be obtained at higher pressures. The curves terminate at voltages beyond which there is significant departure from linearity. The horizontal lines indicate the gas gain required for an adequate signal to noise ratio for events producing a single ion pair when the RMS noise in the preamplifier has the values indicated (see Sec. IV.3.9).

A fundamentally important aspect of gas multiplication is the fluctuation of the charge collected in individual avalanches which is due to statistical variations in

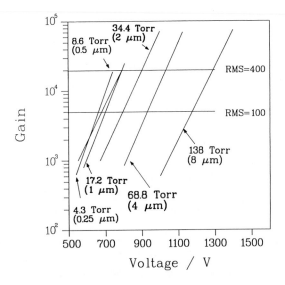

Fig.IV.11 Gain vs applied voltage in a proportional counter at various pressures and simulated site diameters (after Srdoc, 1970). For other details see text.

the distances covered between the ionizing collisions of electrons. This subject has been explored especially by Srdoc and his collaborators (e.g. Srdoc et al., 1987). The distribution of the total charge collected when an event produces n ions in the counter is the n-fold convolution of the single electron spectrum when there is no mutual interference between avalanches.

The spectrum for single electrons can be obtained by UV irradiation of an aluminum surface in the counter. It is well approximated by a *Polya distribution*, p(x).

$$p(x) = x^v e^{-wx}$$ (IV.15)

where v and w are characteristic of the gas. Typical values are v = 0.4-0.6, w = 0.05-0.06 eV^{-1} for alkane gases. The relative variance of this distribution is $(1 + v)^{-1}$. When an energy T is absorbed in the counter gas the relative variance is less by a factor W/T.

Knowledge of the distribution of primary electrons and of the statistics of their multiplication permits calculation of the discrete probability distribution of ions produced when a fixed energy is absorbed in the counter gas. Fig.IV.9 shows the numbers of electrons produced when the 277 eV K_α photons of carbon and 1.5 keV K_α photons of aluminum are absorbed in methane as well as the observed pulse height spectrum. Except for a slight contribution due to instrumental deficiencies, the broadening of the distribution is due to avalanche statistics.

IV.3.5 The Wall Effect

The employment of proportional counters in microdosimetry is based on the assumption that the spatial distribution of energy deposits in the gas-filled cavity is the same as in geometrically similar regions in the surrounding wall material when dimensions are inversely proportional to the density ratio. When both media have the same atomic composition and the density effect on stopping power (Sect.IV.3.3) is neglected this is evidently the case for a particle track that is wholly absorbed in either medium. However in most microdosimetric measurements the dimensions of the simulated sites are less than the range of charged particles that are predominantly produced in the counter wall and usually traverse the counter (these are known as *crossers* as contrasted with *starters*, *stoppers* and *insiders* that begin, end or do both in the cavity, see also section V.3.1). It has been well known that under these conditions the mean energy absorbed per unit mass is the same in wall and gas and formal proof of this was given in a theorem by Fano (1954). However energies deposited in separate events in sites in the wall material can appear simultaneously in a single event in the corresponding gas volume. This has been termed the *wall effect*. Its elimination is

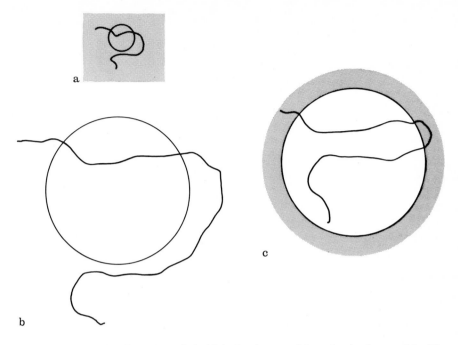

Fig.IV.12 The wall effect: A track in high density **a** and low density **b** materials. The distortion of the track as it penetrates from the high to the low density media is shown at **c**.

the objective in the utilization of *wall-less counters* (Sect. IV.3.7). It has been the subject of an analysis by Kellerer (1971).

A simple example is given in Fig.IV.12 where a curved track is shown at magnifications corresponding to different densities (The diameters differs by a factor of 10 which is very much less than the usual density ratios of wall and gas). The circles surround sites in which a fraction of the track is intercepted if it is entirely located in either medium (a and b). However when after traversing the low density site it reaches the high density medium and is reflected, the low density medium is traversed by another section (c). It is evident that this kind of distortion can occur whenever ionizations are not in a rectilinear array and that it results in a spectrum of fewer and larger events in the counter.

The wall effect is a difficult theoretical problem. Some of its complexity is indicated by the fact that complete knowledge of the distribution of energy transfer points is not sufficient for evaluation. Figure IV.13 shows a configuration of 3 transfer points (A, B and C) in the low density medium. The pattern may be due to two successive ionizations in a track of a high energy particle and the first

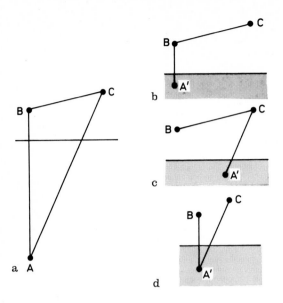

Fig.IV.13 Dependence of the wall effect on the location of the vertex among three energy transfer points (for details see text).

ionization of a delta ray; initial ionizations in 2 spallation tracks; or wide angle scattering of a single particle. In any event, two segments of trajectories, (or those of a single trajectory), must join at a vertex located at one of the points. When the same interaction occurs with A in a high density medium that is replaced by low density medium beyond the line indicated in Fig.IV.13a, the relative position of the points is altered. If the vertex is at either of the points in the low density medium (Fig.IV.13 b,c) their separation is unaffected by the presence of the high density medium and the only difference is an immaterial shift of point A to A' in the high density medium. However if the vertex is at A' (Fig.IV.13d) the separation between points B and C is reduced and they could both be in a gas volume of diameter less than the distance between B and C when A is in the low density medium. Evaluation of the wall effect thus requires not only information on the relative position of the transfer points but also knowledge on the order in which they were produced.

The pattern illustrated in Fig.IV.12 has been termed the *re-entry effect*. It is generally negligible for heavy particles but can be important for electrons, especially those that have an energy below about 1 MeV but a range that is more than the site diameter. Kellerer has estimated that these have a re-entry rate of the order of 20%.

The δ *ray effect* consists in the simultaneous entry of a heavy charged particle and one (or more) of its δ rays as shown in Fig.IV.14. The separate registration of α particles and of δ rays injected by α particles obtained with a wall-less counter as shown in Fig.IV.6, is then vitiated in a walled counter. Kellerer estimates that for 5 MeV protons the frequency of double events is about 15%. However, unlike the re-entry effect for electrons where the energy deposited in the two branches is comparable, the energy deposited by the δ ray is much less than that deposited by the α particle. Consequently, the energy weighted y_D is less affected than y_F. However, if the primary particle is an electron, the energy it deposits in the site can be comparable to, or less than that deposited by its δ ray. Hence wall effects are a major factor in the microdosimetry of electrons (Sect. IV.4.4).

The so called *V effect* occurs when the tracks of two heavy charged particles from a non-elastic nuclear interaction form a V shaped pattern. It is similar to the δ ray effect for heavy particles except that the two tracks make comparable contributions. Hence y_D as well as y_F are substantially overestimated in coincidences caused by the wall effect.

What are termed *wall-less counters* must be contained within a shell and there is a question as to the required relative ratio of the radius R of the shell and radius r of the counter volume. A detailed analysis of the *V effect* which is of major importance in the dosimetry of high energy radiations is given in Appendix IV.7

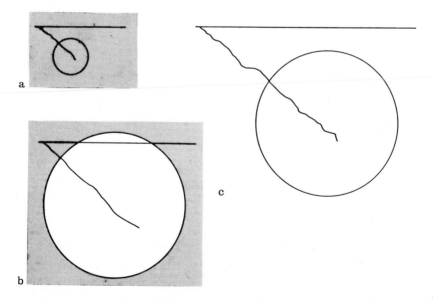

Fig.IV.14 The delta ray effect: A primary particle injects a secondary (delta) electron into a spherical site. **a** and **c** depict high- and low-density media. In **b** because of the difference in density between the cavity and the surrounding material the primary particle and the delta ray appear simultaneously in the cavity.

where it is shown that substantial errors can occur when the ratio of these radii is less than required for the elimination of the wall effect.

IV.3.6 Tissue Equivalent Materials

Although microdosimetry can be utilized in the analysis of radiation deposition in any material, experimental determinations have been restricted to the substances of interest in radiobiology and irradiated media have been considered to have an elemental composition like that of *muscle tissue* as specified by ICRU and given in Table IV.1.

Measurement with homogeneous materials that approximate the bulk composition of tissues (and calculation referring to this composition) does not take account of inhomogeneities which are especially pronounced at the subcellular level (DNA, protein, etc.). The associated local variations of kerma are important only when charged particle ranges are comparable to the distances involved (e.g. electrons of energies less than a few keV). However differences in mass stopping power for particles of any energy can cause deviations between the average values determined and those in various sites in irradiated tissues. This is a limitation of conventional microdosimetry which can be more significant than differences between the atomic compositions of, what are considered to be, tissue equivalent materials, and bulk composition of the tissues simulated. Since the principal components of tissues consist of elements of low atomic numbers, the main source of differences in mass stopping power is the varying hydrogen content.

It has not been found possible to develop a suitable solid material of required composition for counter operations although gelatine based mixtures can be employed in tissue-equivalent ionization chambers (Rossi and Failla, 1956) despite some difficulties (such as loss of moisture). However the high humidity in an enclosure lined with such a gel would drastically interfere with amplification in the counter gas because of negative ion formation in water vapor and propensity to high voltage breakdown. The principal material in the construction of tissue equivalent proportional counters have been conducting plastics in which most of the oxygen of tissue is replaced by carbon. Mixing graphite with a material of high hydrogen content (such as polyethylene) one achieves the proper hydrogen concentration and obtains adequate electrical conductivity and, with the addition of polyamide plastics such as nylon, also the required nitrogen content. Another important advantage is that such plastics can be readily molded and one can manufacture intricate and delicate structures by a technique in which the required shape of the plastic component is removed from a solid aluminum matrix which following the molding procedure is etched away (e.g. with KOH). Although a variety of such plastics were developed early on, the widely used formulation, *A-150*, was developed by Shonka (Shonka et al., 1958). In some instances the air equivalent plastic C-552 is more useful (see below). Somewhat different compositions of these materials have been published. A typical set of data for the composition of both plastics is given in Table IV.1.

A tissue equivalent gas having the elemental composition of ICRU tissue can be made from a mixture of H_2, CH_4, O_2 and N_2 (Rossi and Failla, 1956). Although such a mixture can be employed in ionization chambers, even slight traces of oxygen vitiate proportional counting operation. The *methane based mixture* in Table IV.2 has been employed since the earliest experiments. A *propane based mixture* developed by Srdoc (1970) permits somewhat higher gas gain but is less tissue equivalent. Both mixtures are frequently used. Table IV.2 also gives $(pl)_e$, the product of pressure, p, and distance, l, in the two gases that is equivalent to 1 µm of unit density tissue.

Although liquids are not employed in counter construction they are often convenient as absorbers and scatterers in phantoms into which the counters are placed. Water is often an adequate tissue substitute. More accurate approximations to ICRU tissue have been developed. A simple one designed by Goodman (1969) is given in Table IV.3.

Imperfect matching of the required composition by these solids and gases introduces errors that depend on the type and energy of the radiations. There is also a dependence on the diameter of the simulated site when it is comparable to the range of the charged particles involved because at low energies of uncharged ionizing particles a significant fraction of the charged particles can originate in the gas rather than in the counter wall. The magnitude of the error caused by incorrect composition depends on the criterion by which the spectrum is evaluated.

It is nevertheless possible to provide general statements regarding the effects of lack of true tissue equivalence of A-150 plastic and the counting gases.

In the case of charged particles of energies where nuclear reactions are not significant (usually less than tens of MeV) discrepancies could be caused by differences in mass stopping power which in low atomic number materials are almost exclusively due to differences in hydrogen content. Since this is well matched in the counter gases (which are primarily involved) any errors are likely to be negligible.

For uncharged particles the *energy transfer coefficient* (especially of the wall material) becomes the major factor to be considered. When the primary transfer of photon energy is by the *Compton effect* the very nearly equal electron densities obviate significant errors. Below about 50 KeV the *photoelectric effect* becomes increasingly important and there may be discrepancies of about 30% at 10 keV in A-150. C-552 which causes somewhat (10%) poorer results at high photon energies is a good substitute between 50 and even 1 keV. However at energies below about 10 KeV the response of counters, as usually employed, is mostly due to electrons produced in the gas. This may cause a decrease of kerma of up to 20%

down to photon energies of 0.5 keV in the methane based gas and somewhat more in the propane based gas. Differences in the shape of the spectra are less important.

Except for narrow resonances in heavier elements the bulk of the absorbed dose due to neutrons of energies less than 10 MeV is due to hydrogen and at low energies also due to nitrogen. The proportion of both of these elements is well approximated in A-150 and adequately in the two counting gases. At high neutron energies elastic and inelastic collisions with carbon and oxygen become important and lack of tissue equivalence can cause appreciable (>20%) errors. Table IV.4 gives values of y_D for various substances as determined on the basis of theoretical spectra (Green et al, 1990).

Table IV.1
Elemental Compositions (Percent by weight) of Tissue Equivalent Solids

	H	C	N	O	F	Si	Ca	F
ICRU Tissue	10.1	11.1	2.6	76.2	---	---	---	---
A-150 Plastic (Tissue Eq.)	10.2	76.8	3.6	5.9	1.7	---	1.8	---
C-552 Plastic (Air Eq.)		2.5	50.2	---	0.4	46.2	0.4	---	---

Table IV.2
Elemental Composition (percent by weight), components (percent by partial pressure) of tissue equivalent gases and (pl)/kP.cm*

	Elemental Composition				Components				$(pl)_c$
	H	C	N	O	CH_4	C_3H_8	CO_2	N_2	
Methane based	10.2	45.6	3.5	40.7	64.4	0	32.5	3.1	74.6
Propane based	10.3	56.9	3.5	29.3	0	55	39.6	5.4	43.7

*Product of pressure (torr) and distance (cm) in the gas that is equivalent to 1 micrometer of tissue.

Table IV.3

Elemental Composition (percent by weight) and Components (percent by weight of tissue equivalent liquids)

	Elemental Composition				Components		
	H	C	N	O	H_2O	Glycerol	Urea
Water	11.1	0	0	88.9	100	0	0
T.E. Liquid	10.2	12	3.6	74.2	65.6	26.8	7.6

Table IV.4

Calculated values of y_D in various materials Neutron Energy, E_n, (MeV)

	E_N			
	5	10	15	19.5
Al50	53.8	73.4	101.1	124.7
Water	46.8	55.0	58.9	47.8
Adipose Tissue	51.9	68.3	90.9	109.4
Brain	48.2	57.9	65.9	62.4
Kidney	49.3	60.4	69.5	66.3
Muscle	49.3	60.3	68.9	64.7
Liver	53.8	60.0	69.7	68.0

IV.3.7 Counter Designs

In the following, examples of counter design dimensions given in the metric system are approximate because the original dimensions are in the English System.

The spherical walled proportional counter design adopted in the early period of microdosimetry has been retained with minimal changes and it remains to be the only version that is currently available commercially[15]. The design has been employed in counters having (inside) diameters ranging from 0.6 to 30 cm. This corresponds to counting rates that differ by a factor of 2500. Fig.IV.15 shows a 10 cm version. The helix provides a more uniform electric field along the anode. Both of these structures are made from 60 μm stainless steel wire. The diameter of the helix is 6.4 mm or about 1/16 of the cavity diameter. The pitch of the helix (turns per unit length) is 0.3 mm^{-1} (i.e. about 2 turns per helix diameter). In larger and smaller counters the relative dimensions are generally maintained but appear not to be critical within a factor of 2 or 3. Other designs do not incorporate a helix but employ field shaping electrodes in the vicinity of the poles of the counter (Benjamin et al., 1968). This simplifies construction but results in somewhat poorer resolution. A small port that is plugged in this illustration serves as an opening for the insertion of a calibration source.

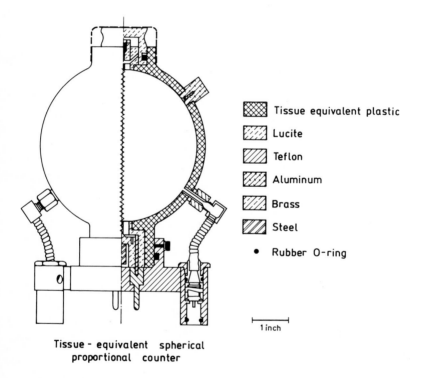

Tissue equivalent plastic
Lucite
Teflon
Aluminum
Brass
Steel
● Rubber O-ring

1 inch

Tissue - equivalent   spherical
proportional   counter

Fig.IV.15 Typical walled proportional counter employed in microdosimetry.

---

[15] Except for replacement of the helix with field shaping electrodes.

In early designs of wall-less counters the collecting volume was largely defined by lines of force (Rossi, 1967). However, recent designs employ grids of tissue equivalent plastic because this minimizes diffusion of electrons which are collected more rapidly. Fig.IV.16 shows a version in which ribs formed in the manner of latitude circles on a globe are arranged in 3 mutually orthogonal sets. The ribs are about 100 μm wide and 200 μm deep and more than 90 percent of incident particles traverse the structure which is typically at a potential of -100 V with respect to the grounded wall of a "wall-less" counter which is illustrated in Fig.IV.17. This is a 6.4 mm counter in a 10 cm shell.

In a counter which is employed in the microdosimetry of low energy X radiation stainless steel electrodes were replaced by similar structures fabricated from tissue equivalent plastic (Fig.IV.18). The helix was replaced by two counter-current molded tissue equivalent helices. The pitches of the two helices are about 5 and 5.5 cm^{-1} which avoids co-linear nodes that may cause a directional response. The anode is a (carefully selected) human hair coated with colloidal graphite. The helix and the shell counter are made from C-552 plastic which is more rigid and has a more appropriate response to low energy photons.

Adoption of cylindrical rather than spherical counters results in considerable simplification of designs with smaller dimensions and consequent reduction of counting rate. The usual geometry is one of a cylinder of height equal to its diameter. This frequently results in a distribution of intercepted random particle trajectories that is nearly the same as that in a sphere (Section IV.2). Cylindrical counters may therefore be employed when the angular distribution of particles is not far from isotropic.

1 cm

Fig.IV.16 Spherical grid defining the volume of a wall-less counter.

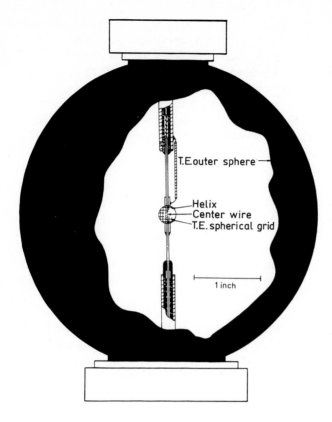

Fig.IV.17 Typical wall-less counter.

Walled cylindrical counters having dimensions of the order of 1 mm have been successfully operated. Figures IV.19a and IV.19b show the general arrangement and detail of the center of a 1 mm counter. The diameter of the anode is 25 μm. The slits in the central wall permit passage of a particles or photons employed in calibration utilizing sources that can be externally rotated into on and off position. The counting rate of this device is approximately 10^{-5} that of the largest (30 cm) spherical type. This not only permits measurement at relatively high dose rates but also makes it possible to determine the microdosimetric spectrum of neutrons that are produced by pulsed beams of high energy x rays. The interaction in tissue can result in an absorbed dose rate of protons that is typically 1% of that of the electrons and the relative fluences differ by a factor of the order of 10^4. If in a small counter only few electrons (~10) appear in a pulse, the occasional pulses due to the simultaneous presence of protons are much larger with the result that their microdosimetric spectrum can be determined in a statistical analysis. Another advantage of small counters is that they can be employed in phantom measurements.

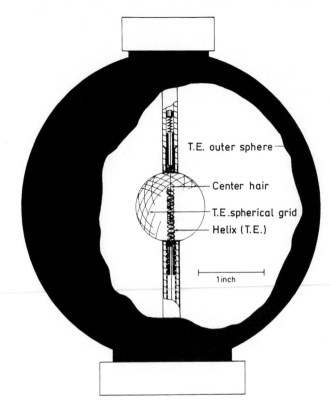

Fig.IV.18 Non-metallic wall-less counter.

For low dose rate measurements in phantoms the design in Fig.IV.20 provides for a higher counting rate in a compact device. It incorporates 296 cylindrical counters of 3 mm height and diameter that operate in parallel in a 2.5 cm enclosure. Because of the large total surface of the miniature counters the counting rate is about 6 times larger than that of a 2.5 cm diameter spherical counter. In measurements within a phantom, the average density of about 0.3 reduces the perturbation of the radiation field that would be caused by the large cavity of the equivalent single counter and it happens to be especially suitable for measurements in a lung phantom. The resolution expressed in FWHM (Sect.IV.3.11) is about 38% for 1.5 keV x rays compared with 30-35% which is optimal for single counters. Disadvantages are difficult construction of this elaborate design (Kliauga et al., 1989) and the limitation that charged particles must not traverse more than one of the counters (which would increase the apparent LET) because they penetrate the intervening septa. It is estimated that this restricts the use of the instrument to neutron energies below 14 MeV.

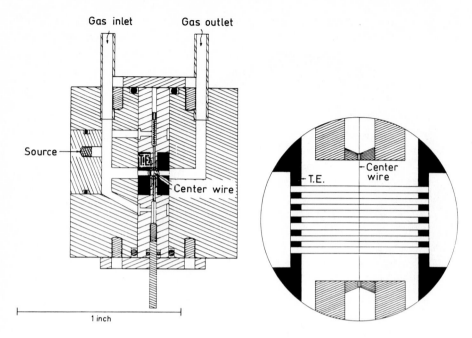

Fig.IV.19 a) 1-mm counter; b) enlarged detail of the central region.

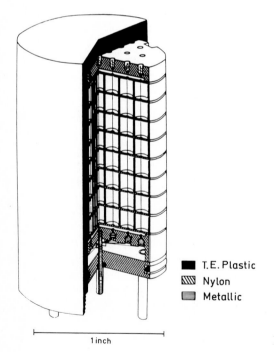

Fig.IV.20 Multi-element counter.

Wall-less cylindrical counters have been employed in the microdosimetry of charged particles and especially that of their delta rays. The sensitive volume is defined by a helix and field tubes which intercept ions outside of the sensitive volume and are at an appropriate potential to maintain radial lines of force. Experimental investigations have shown that a substantially poorer resolution is obtained in an alternate design where the total length of wire in the helix is divided in a circular array of wires that surround the anode and are parallel to it. Figure IV.21 shows a wall-less cylindrical counter which was placed in a large tank (a cylindrical cavity of 2.5 m length and 0.6 m diameter) in experiments to investigate the delta ray pattern around heavy ions of energies of several GeV. (see section IV.4.5).

Unusual designs have been developed for special purposes. Figure IV.22 shows a dual counter employed in the variance-covariance technique (see section IV.5.2).

Detailed information on counter manufacturing techniques and the similar methods employed in the construction of tissue equivalent ionization chambers is given in appendices of ICRU reports 26 (1977) and 36 (1983).

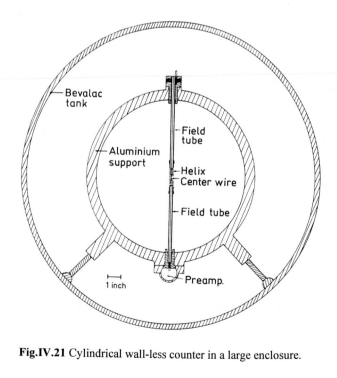

Fig.IV.21 Cylindrical wall-less counter in a large enclosure.

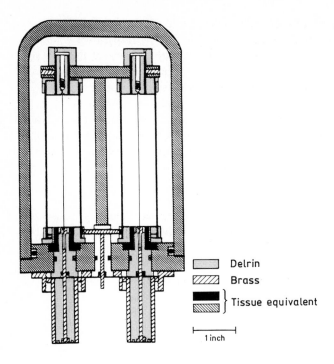

Delrin
Brass
} Tissue equivalent

├──────┤
1 inch

Fig.IV.22 Dual counter employed in the variance-covariance method.

IV.3.8 Gas Supply

The quality of microdosimetric measurements depends critically on the gas in proportional counters. It is necessary that the gas have the correct atomic composition with especial care being taken to avoid contamination by electronegative gases such as oxygen or water vapor. Srdoc and Sliepcevic (1963) have shown that admixtures of oxygen even in the ppm range can cause erratic counter performance. An equally important requirement is constancy of gas pressure on which amplification depends exponentially. The mass of gas in a counter is typically of the order of 1 mg and tissue equivalent plastic adsorbs gases introduced into the counter, releases other gases (including oxygen) to which it was previously exposed and even permits diffusion of gases through thin layers. These processes are particularly important at the unfavorable surface to volume ratios of small counters.

The best way of ensuring constant composition and pressure is to employ a properly designed gas flow system. However, because of the complication and physical restrictions involved, counters are frequently employed in a sealed mode in which they may operate satisfactorily for days or even weeks provided the seals are adequate and measures are taken to minimize changes in the gas composition in the counter. These include initial evacuation for at least several hours, with the counter preferably at elevated temperature, followed by a constant gas flow at the pressure at which the counter is to be sealed several hours later. Some counters are operated in an aluminum housing with tissue equivalent gas at the desired pressure both inside and outside of the tissue equivalent wall.

Methane and propane based TE gases are both available commercially. They may also be prepared by mixing the component gases according to partial pressure in a large volume from which they are connected to the counter. In order to achieve the proper mixture mechanical agitation is advisable because diffusion between successive gas layers can be a slow process.

The schematic arrangement and a photograph of a flow system are shown in Fig.IV.23 and Fig.IV.24. Constancy of pressure is achieved by employment of a Cartesian manostat. The slightly different gas pressures are measured at the inflow and the outflow of the counter with the pressure taken to be the average of the readings. The flow rate is typically a few ml/s at STP.

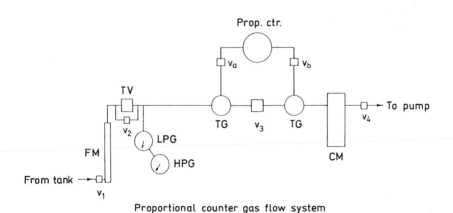

Proportional counter gas flow system

Fig.IV.23 Schematic view of a typical gas flow system. v_1, v_2, v_3 and v_4: shut-off valves; v_a and v_b: spring-loaded valves in counter (shown in Fig.IV.15); FM: flow meter; LPG: pressure gauge 0-50 Torr; HPG: Pressure gauge 0-800 Torr; TV: throttle valve; TG: thermo-couple gauges; CM: Cartesian pressure regulator.

IV.3.9 Electronics

The fast rise time signal produced by free electrons is very small compared to the signal resulting from the drift of positive ions toward the cathode. The avalanche consisting of 10^4-10^7 ion pairs is formed in a narrow zone around the anode wire, and it is the drift of the positive ion cloud in the strong electric field near the

Fig.IV.24 View of the counter and the gas flow system.

anode which produces the electric signal proportional to the total number of charges in the avalanche.

Two additional sources of measurement distortion should be considered. High voltage ripple and/or drift causes change of gas multiplication (typically, a 5% increase of the counter voltage doubles the pulse size in a proportional counter). Fortunately, most of the modern commercially produced power supplies have a voltage stability close to 0.1-0.2% and no observable voltage ripple. Another possible source of spectrum distortion related to the high voltage are spurious pulses caused by the electric breakdowns across the surface of the insulator separating the anode from the counter shell or the helix. This can be eliminated by keeping the insulators clean and it is usually also minimized by increasing the leakage path such as "undercutting" (slight reduction of the diameter of cylindrical insulators near their ends) or grooves (on the flat surfaces of the insulator). Guard rings (i.e. intermediate electrodes at a potential that is essentially that of the anode) are rarely required but they are employed in some cases as in the dual counter shown in Fig.IV.22 because even minor fluctuations are important in the variance-covariance technique (see sect. IV.5.2).

In the most usual arrangement the counter shell and the preamplifier are at ground potential and the helix at intermediate voltage. The anode is at high voltage and coupled to the preamplifier input via a ~10-100 pF capacitor. The breakdown voltage rating of the coupling capacitor must be 2 or 3 times higher than the highest anode voltage employed. Only high quality commercial capacitors do not produce spurious pulses at the voltages usually applied in proportional counter operation (about 500 to 5000 V).

The principal limitation to useful electronic amplification is the preamplifier noise. This is usually measured in the RMS (the root mean square, i.e. the standard deviation) of the number of electrons at the preamplifier input, although the shape of this temporal distribution is also of some importance. The complex subject of electronic noise has been analyzed in detail by Radeka (1968). Its practical implications to microdosimetry have been discussed by Srdoc (1970).

When the detection threshold is at 5 x RMS most of the noise is suppressed with only about 1 cps (counts per second) as typical counting rate due to background noise. This is sufficiently low in most measurements where the counting rate due to radiation is likely to be larger by a factor of 100 or more. A possible exception can occur at radiation protection surveys involving low LET radiation which may produce as little as a few or less than 1 cps at pulse heights that can be only moderately above the noise level. In this case a bias of 10 x RMS may be required.

Since the noise and the signal are subject to the same electronic amplification it is necessary to increase the gas multiplication if the signal is below the lower discriminator level (LDL) defined by the noise. Maximum gas gain is required for single ionizations, i.e. one ion pair or one electron per event in the sensitive gas volume. As stated in Sect.IV.3.4 this approaches a Polya distribution and the registration of most of the counts requires that the average signal be about 10 times as large as the discriminator level. It follows that the gas gain should be 50 x RMS if most of the pulses due to single ions are to be collected. This is necessary when the lineal energy spectrum of hard gamma radiation is measured in sites of diameters that are about one μm or less.

The RMS of most of the commercially available preamplifiers is near 400 electrons which according to the criterion established by Srdoc (1970) requires a gas gain of 2×10^4. As can be seen from Fig.IV.11 this is near or beyond the practical range of gas multiplication in small sites. A superior amplifier design by Radeka (1968) employs a field-effect emission transistor. It is a charge sensitive device and operates best when the anode is at ground potential; if this is not feasible or practical, the FET preamplifier should be coupled to the anode via a capacitor of several pF. The pre-amplifier is therefore best integrated into the counter structure and RMS values as low as 100 electrons can be achieved.

The pulse emerging from the preamplifier is shaped in the linear amplifier over a period that is sometimes made adjustable because its optimum value depends on the parameters of counter operation. It is typically of the order of 0.1 μs and imposes a limit of the order of 10^4-10^5 cps for most applications. The dynamic range of linear amplifiers is generally 10^2-10^3 and overloading causes saturation followed by a long recovery time during which linearity is lost and even false pulses may appear. It is essential that the output of the linear amplifier be monitored with an oscilloscope.

Since the range of pulse heights can exceed 10^5 it is frequently necessary to determine microdosimetric spectra in two or more sections. Other methods employ logarithmic amplifiers or several linear amplifiers operated in parallel and at different gains.

The amplified pulses are conducted to a multichannel analyzer which is usually connected to a computer for further processing of data that includes determination of the f(y) spectrum on the basis of the calibration spectrum, normalization of the f(y) spectrum, computation of the normalized d(y) spectrum and its moments, graphics and printing.

IV.3.10 Calibration

It is a major advantage that proportional counters can be calibrated *absolutely*, in the sense that comparison to standard instruments is not needed, because accurately known energy can be imparted to the counter gas. In the microdosimetry of neutrons a rather good calibration is even obtained without exposing the counter to another radiation source because of a phenomenon known as the *proton edge*. This is due to the fact that in tissue the maximum LET of protons (as exhibited by the Bragg curve) is near 95 keV/μm (the exact value in TE materials is within 5% of 95 kev/μm when the proton energy is about 100 keV). Neutrons impart energy to tissue primarily via proton recoils of maximum energy equal to the neutron energy. Consequently, the f(y) spectrum for neutrons of energy in excess of 100 keV exhibits a sharp drop which is due to the fact that (apart from straggling) the maximum energy, ε, any proton can deposit is 100d keV when a proton passes along a major diameter d(μm). Since the lineal energy is defined by $\varepsilon / \bar{l}$, and $\bar{l} = d / 1.5$ for a sphere, the proton edge occurs at a lineal energy near 150 keV/μm in spherical counters.

Two factors contributing to the distinctness of the proton edge are the zero derivative of the Bragg curve at its maximum, and that the most probable chord length in a spherical cavity is equal to the diameter. There are, however, several factors that limit the accuracy of calibrations based on the proton edge. One is that in traversing the counter diameter a proton cannot maintain maximum LET especially when the simulated diameter exceeds 1 μm; others are that straggling obscures the precise position of the edge and that at high neutron energies heavier recoiling nuclei produce lineal energy around and above the edge.

A method of calibration suitable for high-LET radiation employs a source of α particles (e.g. ^{244}Cm, ^{243}Am, etc.) that is located near the inside surface of the counter and it is based on the known energy and hence known LET, L, of the α particles. It is essential that the source be "thin" i.e. exhibit no significant absorption in the active portion. This can be determined with solid state (barrier) detectors. In most practical instances the maximum of the bell-shaped pulse height distribution can (in a spherical counter) be considered to correspond to a lineal energy given by

$$y = \frac{3L}{2}$$ (IV.16)

where d is the equivalent diameter in unit density tissue.

For calibrations at lower energies, x ray sources can be employed that emit accurately known K_α radiation that is totally absorbed in a gas in photoelectric absorption. The sources may be radioactive (e.g. ^{55}Fe) or a miniature x ray tube that fits into an opening of the counter wall and emits substantially monoenergetic x rays (Srdoc and Clark, 1970). Typical targets are carbon and aluminum with K_α lines of 227 eV and 1.49 keV, respectively. Major advantages of this technique are the very accurately known photon energy and substantial independence on the configuration of sources or absorbing materials. It is, however, necessary that the range of the photoelectrons is significantly less than the counter diameter in order that the bell-shaped curve corresponding to the full range of photoelectrons be well separated from smaller pulses produced by electrons that expend only part of their energy in the counter (although Compton electrons have lower energies, their number is negligible).

A somewhat more involved calibrating procedure consists in mixing a brief burst of ^{37}Ar with the TE gas in counters operating under flow conditions. It has the advantage of providing two lines (270 eV and 2.62 keV) which permits a check on the over-all linearity of the measuring system.

IV.3.11 Resolution

When charged particles deposit equal energy in a proportional counter a range of pulse heights is observed because of various phenomena considered in previous sections. The width of this distribution is usually specified in terms of the "percent resolution", R, as $(FWHM/x_{max})100$, where FWHM (the *full width at half maximum*) is defined as the width of the distribution between the points where it reaches half of the maximum value x_{max}. In a normal (Gaussian) distribution, which is often approximated in experimentally obtained spectra

$$R_{FWHM} = 2.354 \frac{\sigma}{\bar{x}} 100(\%) \qquad\qquad (IV.17)$$

where σ^2 is the variance and $\sigma / \bar{x}$ is the relative standard deviation of the distribution.

The resolution refers to microdosimetric events of equal magnitude and does, therefore, not include statistical fluctuations that are among the objectives of measurement (as considered in section IV.2). It is caused by two classes of additional fluctuations.

The first class involves two phenomena that are inherent and essentially unavoidable in proportional counter operation. They determine what is termed the *theoretical resolution*, R_{th}. One of these is the relative variance $(W/T)F$ in the number of ions produced when the energy T is absorbed from particles that expend the energy W in the production of an ion pair, with F being the Fano factor (see sect. IV.3.3). The other fluctuation is due to the statistics of gas multiplication. As stated in Sect.IV.3.4, its relative variance is $(W/T)(v+1)^{-1}$ where v is the exponent in the Polya distribution. The relative variances are additive and the theoretical resolution is

$$R_{th}(\%) = 2.345 \sqrt{\frac{W}{T}\left[F + (v + 1)^{-1}\right]} \, 100 \qquad\qquad (IV.18)$$

The second class of phenomena determines the instrumental resolution, R_{inst}, which depends on a large number of imperfections. They include fluctuations of the counter potential and of the gas pressure, non-uniformity of the electric field in the counter and preamplifier noise. In a properly designed and operated counter R_{inst} is usually no more than 10% of R_{th} and since

$$R = \sqrt{R^2_{th} + R^2_{inst}} \qquad\qquad (IV.19)$$

instrumental deficiencies should make only a minor contribution to R. Experimental verification of this condition is best carried out by employing the soft x-ray sources discussed in Sect. IV.3.10 because unlike collimated α ray sources they may be considered to produce events that are not subject to intrinsic fluctuations due to straggling. Figure IV.25 shows the relation between T (the photon energy) and R_{th}. This may be compared with the measured value of R, and R_{inst} can be determined on the basis of Eq(IV.19).

The relative variance of a measured lineal energy distribution is

$$V_T = \frac{y_D}{y_F} + R^2 - 1 \qquad\qquad (IV.20)$$

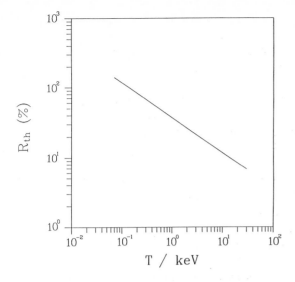

Fig.IV.25 R_{th} is the theoretical resolution. Within the accuracy of the graph (about 5%) it applies to both methane and propane-based tissue equivalent gases, and to photons as well as electrons of energy T.

Hence if R has been determined, an appropriate correction can be made in the calculation of y_D which is an important microdosimetric quantity. Thus

$$y_D = (V_T + 1 - R^2)y_F \qquad (IV.21)$$

The correction is usually small and frequently neglected.

IV.4 Measured Distributions of Lineal Energy

IV.4.1 General Comments

There exists a sizable literature covering the results of microdosimetric measurements, and several hundred spectra have been published. With some exceptions (e.g. ^{60}Co γ radiation or 14 MeV neutrons) only single, or few, measurements have been carried out for the same radiation quality and the same site diameter. Comparisons of the results must, therefore, be limited. To the extent that they are possible, a standard deviation of about 10% would seem likely when the measured portions of spectra are compared. One of the principal difficulties is limited accuracy, or absence, of the spectrum of small values of lineal energy.

This is especially important for low-LET radiation. Many determinations are restricted to values above several hundreds eV/μm and various extrapolation procedures have been employed to estimate the shape of the lower portion of the spectrum. This can cause considerable uncertainties in the determination of the event frequency average, y_F. However there are then often only minor uncertainties in the dose average, y_D. As already noted, y_F is proportional to the reciprocal of the *event frequency* Φ^* which implicitly determines the fraction of sites that at a given absorbed dose experience no events. y_D is more closely related to the probability of biological effects.

Measurements with walled counters effectively eliminate contributions of uncorrelated delta radiation (Sect. IV.3.5). However, even when wall-less counters are employed in measurements involving ions or neutrons the small pulses due to delta rays are frequently ignored with consequent appreciable errors in the determination of y_F and hence of the event frequency.

As shown in section IV.3.11 there is extensive information on the factors causing limited resolution in proportional counter measurement and it is therefore possible to correct measured pulse height distributions and to eliminate this source of error. This has rarely been attempted. One major exception is the work of Srdoc and his collaborators (1987) who derived the varying number of ions produced by monoenergetic low energy x rays and separately calculated the magnitude of this fluctuation, the statistics of avalanche formation and the additional instrumental errors. One reason why this kind of analysis is infrequently made is that in many spectra these factors cause modifications that are small compared with the effect of straggling.

Experimentally determined spectra differ from calculated ones if the latter do not take account of straggling. As expected, the omission results in sharper distributions. Theoretical treatments that include straggling agree well but involve more complex calculations. The utilization of such information requires knowledge of the energy spectrum of radiations which can often be difficult to evaluate; measurements of lineal energy spectra may then be essential. This is especially common in radiation protection which frequently deals with radiation energies and even radiation types that are quite different from those emitted by the primary source. It can also be the case in therapy where collimators and absorbers cause substantial changes of the energy spectrum of the radiation originating at the source.

Lineal energy distributions usually extend over a wide range of y and in that case they are generally presented in graphs in which y is marked on the abscissa on a logarithmic scale and the products yf(y) or yd(y) on the ordinate in a linear scale interval of y. Between two values of y the area under such curves represents the fraction of events or of absorbed dose due to lineal energies in this interval (see section II.3).

In addition to the averages y_F and y_D the quantity y^* (see section VI.2.1) is frequently evaluated in studies of neutron therapy beams because it represents an approximate measure of the RBE.

Several investigations of mixed radiations have employed the *time of flight* technique in which "electronic gates" are applied to the counter permitting separation of the spectra due to particles that are fast (e.g. photons) or slower (e.g. neutrons) when the output of pulsed generators is studied.

The sheer volume of the literature necessitates the selection of typical examples. Additional ones that relate specifically to radiation protection and to radiotherapy are presented in Sect. VI.2 and VI.3, respectively. Further references can be found in the articles quoted.

IV.4.2 Neutrons

From the beginning microdosimetric determinations have concentrated on neutrons. Next to photons they are the most important radiation involved in protection as well as in therapy. The energy dependence of biological effectiveness is far more important for neutrons than for photons and the RBE of neutrons relative to photons can exceed 100.

Most of the data have been obtained with walled counters although in several instances wall-less counters were employed. Comparisons of the data obtained with the two types indicate that even at fairly high energies differences are small. An example (Menzel et al, 1983) is shown in Fig.IV.26. In these spectra of d(y) the major difference which involves the number of small pulses is not obvious because they contribute little to the absorbed dose.

Gamma radiation is invariably associated with neutrons. Even when neutrons are generated in nuclear reactions that do not involve prompt or delayed gamma ray emission γ photons are produced as a result of neutron capture in the environment of accelerators (e.g. the shielding) and nearly all accelerators emit X radiations generated internally by high energy electrons. Although the photon kerma may be small compared with the neutron kerma the counting rate due to electrons can be substantial because for equal kerma the number of electrons is up to 100 times larger. The subtraction of electron pulses, which is a major concern when gamma radiation is actually emitted in neutron production (e.g. in fission), has been made by various techniques including utilization of a ^{60}Co spectrum. Various techniques that do not only separate counts due to γ radiation but also those to the protons and heavy recoils produced by neutrons are described by Booz and Fidorra (1981). An example of a mixed field of gamma and neutron radiation is that due to ^{252}Cf (Fig.IV.27).

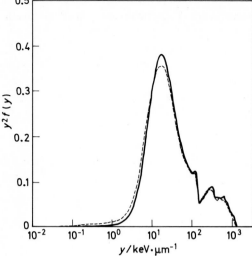

Fig.IV.26 Dose distribution of lineal energy due to 8.5-MeV neutrons as determined by a walled (solid line) and a wall-less (dashed line) counter (after Menzel et al, 1983).

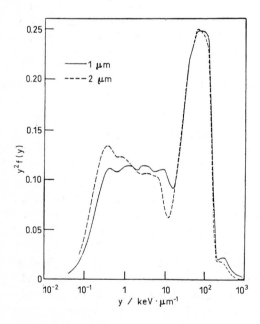

Fig.IV.27 Dose distribution, yd(y), of lineal energy due to ^{252}Cf for site diameters 1 and 2 μm (after Dicello et al, 1972).

The change of the lineal energy spectrum of neutrons with energy and site diameter has been demonstrated in a number of publications beginning with the first study (Rosenzweig and Rossi, 1959). A more recent example is shown in Figs.IV.28a and IV.28b and illustrates the principal characteristics of neutron spectra (Srdoc et al, 1981).

At a given diameter, d = 1 μm, when the neutron energy is 0.22 MeV, the energy-weighted average proton energy of about 0.15 MeV is spent near the maximum

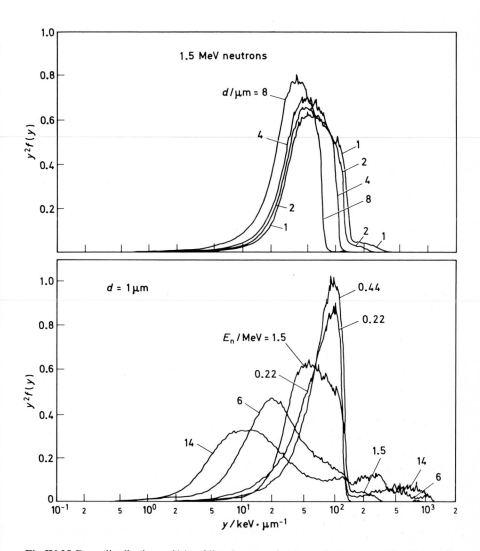

Fig.IV.28 Dose distribution, yd(y), of lineal energy due to neutrons measured with a wall-less counter. a) Energy 1.5MeV, site diameters between 1 and 8 μm; b) Site diameter 1 μm, energies between 0.22 and 14 MeV (after Srdoc et al, 1981).

LET of about 100 keV/μm, which corresponds to a proton energy near 100 keV. Consequently, the proton spectrum is narrow. As the neutron energy (and hence the mean proton energy) is increased the proton spectrum extends to much lower values of y. Beyond the proton edge at y = 150 keV/μm the contribution of heavy recoils (C, O and N nuclei) is evident and it increases in importance with increasing neutron energy.

At a given neutron energy (E_n=1.5 MeV) an increase of d causes narrowing of the distribution because of reduction of straggling but the *proton edge* shifts downward because at the larger sizes the maximum energy that protons can expend in the cavity is less than the product of the maximum LET and the diameter. The contribution of heavy recoils is entirely within the proton spectrum if d = 8 μm. At this diameter y_{max}~ 75 keV/μm and the maximum energy deposited by traversing protons, $(2d/3)y_{max}$ = 400 keV, is more than that deposited by maximum energy recoils even if these originate and terminate in the gas volume (~0.25 x 1.5 MeV = 375 keV).

When neutrons are produced in pulses by accelerators in which the ion current is discontinuous it is possible to utilize time-of-flight techniques in which only events due to neutrons arriving within the narrow time interval are registered. This method has been employed to separate even the spectra due to neutrons of different energy (Randers-Pehrson et al, 1983). It requires rather complex electronics but could ultimately permit measurement of y-distributions of substantially monoenergetic neutrons among those emitted by polyenergetic sources besides virtual elimination of pulses due to gamma radiation.

Because of the substantial depth doses required, radiotherapy usually utilizes neutrons of high energy. Examples of spectra are given in section VI.2.2a.

At the other extreme of neutron energy spectrum, the microdosimetry of slow or thermal neutrons has been investigated. The complexity of capture reactions (especially by hydrogen in tissue) makes the results strongly dependent on irradiation geometry and atomic composition including that of the proportional counter itself. Because of the particular importance of thermal neutrons in radiation protection their spectra have been determined in counters designed to determine the operational quantities and especially H^* (see sect.VI.3). Figure IV.29 shows the d(y) spectrum registered by such a device as well as the pulse height distribution which is essentially proportional to f(y) (Booz, 1984). A practical presentation of the data differs from that at higher neutron energies because a logarithmic scale is appropriate for the ordinate and a linear one for the abscissa. Multiplication by the quality factor shows that the vastly greater number of electrons makes only a small contribution to the total dose equivalent. Further details are given in Section VI.3.

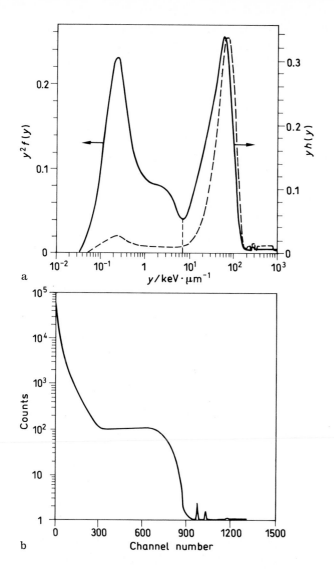

Fig.IV.29 (a and b) Thermal neutrons. **a** dose distribution yd(y) of lineal energy (solid line) and the distribution yh(y) in which y is weighted by the quality factor Q; **b** pulse height spectrum, which is essentially proportional to f(y) (after Booz, 1984).

IV.4.3 Photons

Lineal energy distributions have been determined for a wide range of photon energies. The data of interest are restricted to energies above about 10 keV because below this value the range of photoelectrons (essentially the only secondaries) becomes comparable to the smallest sites that can be reasonably simulated in proportional counters. Hence below this limit the y spectrum due to monoenergetic photons approaches a relatively narrow peak which makes low energy x rays (such as the K_α lines of C and Al) very useful for calibration. As explained in sect. IV.3.11, the width of this peak (the resolution) is entirely due to experimental limitations (fluctuations in the number of ions produced, statistics of avalanche formation and instrumental deficiencies) because such x rays can be considered to be strictly monoenergetic. As mentioned in sect. IV.4.1, Srdoc et al. (1987) have performed a detailed analysis of the factors responsible for the pulse height spectrum from line sources. Corrections for finite resolution are unnecessary in photon measurements at higher energies because of the overwhelming dominance of straggling except in small sites where the spectrum produced by few ion pairs is appreciably affected by counter resolution.

Wall effects are clearly significant with photon beams of moderate and high energy. In terms of y_F Braby and Ellet (1971) have shown them to range from about 10 to nearly 30 percent for unfiltered and filtered x ray beams between 60 and 250 KV and Amols and Kliauga (1985) reported a value of about 30% for 10 MeV x rays.

Lineal energy spectra for a range of photon energies (12 to 1250 keV) and site diameters (0.24 to 7.7 μm) were published by Kliauga and Dvorak (1978). Fig.IV.30 shows the variation of d(y) with energy for a 1 μm site diameter and Fig.IV.31 shows the variation with site diameter for 25.4 keV photons. These data were obtained with a wall-less counter. The extensive volume of data obtained in this study permitted construction of reasonably accurate continuous curves for y_F and y_D for the range of energies and diameters investigated. It was also possible to obtain an understanding of the relative importance of photoelectrons and Compton electrons on the basis of a theorem by Kellerer which states that (see Eq.IV.2)

$$1 / y_F = 1 / \overline{L_T} + 2d / 3\overline{E} \qquad (IV.22)$$

where y_F is the average lineal energy produced in a spherical site of diameter d by particles of average energy $\overline{E}$ having a track average LET equal to $\overline{L_T} = \overline{E} / \overline{R}$ with $\overline{R}$ being the mean range. The formula applies for particles that stop or originate in the cavity but it is assumed that the tracks are straight line segments. The assumption is sufficiently well met so that a plot of $1/y_F$ vs d results in rather straight lines of slope $2 / 3\overline{E}$ with an intercept $1 / \overline{L_T}$ at the ordinate. If $\overline{L}$ is obtained by this procedure, a plot (Fig.IV.32) illustrates the relative contributions of the two sources of electrons and their average LET as a function of photon energy.

There are numerous determinations of y spectra of ^{60}Co γ rays. A typical example is shown in Fig.IV.33 (Coppola as quoted by Booz, 1976). These data obtained with a cylindrical wall-less counter extend down to d = 0.1 μm. Although there

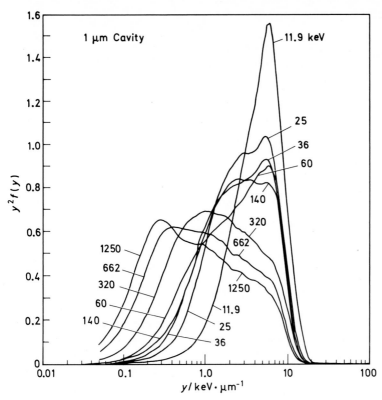

Fig.IV.30 Dose distributions yd(y) due to photons. Site diameter 1 μm; energies between 11.9 and 1250 keV (after Kliauga and Dvorak, 1978).

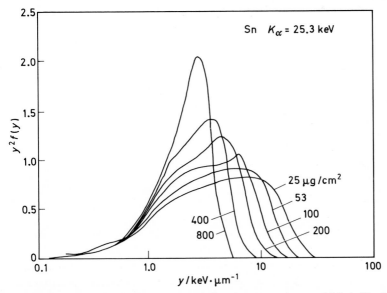

Fig.IV.31 Dose distributions yd(y) of lineal energy due to 25.3 keV photons. Site diameters between 0.25 and 8 μm (after Kliauga and Dvorak, 1978).

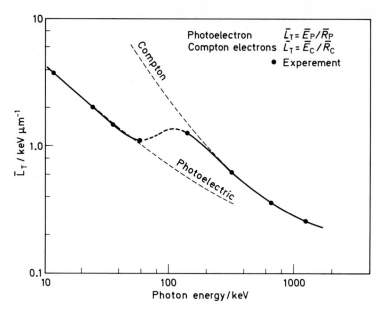

Fig.IV.32 Track averaged LET, $\bar{L}_T$, as a function of photon energy showing the contribution by the photoelectric and Compton effects (after Kliauga and Dvorak, 1978).

are theoretical reasons why the resolution of the counter should be poorer at such diameters (Sect. IV.3.4) this is likely to be masked by the more important straggling.

Microdosimetric spectra for 10 MV x rays differ little from those of ^{60}Co γ rays (Amols and Kliauga, 1985) and there is hardly any difference between the spectra of 42 MV x rays and 8 MeV electrons (Lindborg, 1976). At such energies no significant change of the spectrum was observed in depths exceeding 15 cm in tissue equivalent phantoms. Slight contributions at large y (due to nuclear reactions) were detected.

IV.4.4 Electrons

The microdosimetry of electrons involves substantial difficulties. Most of the enclosures in which proportional counters must operate stop electrons of energies less than 1 MeV. Studies with internally located β ray sources are of no general interest because electrons, unlike α particles, are strongly scattered and the lineal energy spectrum obtained depends on the geometry of both the source and the

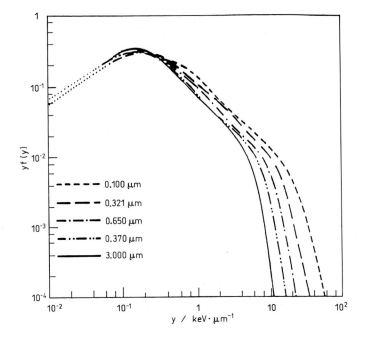

Fig.IV.33 Frequency distribution yf(y) due to ^{60}Co gamma radiation. Wall-less cylindrical counter. Site diameter (and equal height) between 0.1 and 3 µm (after Coppola and Booz, 1975).

counter. An exception are gases that emit low energy β radiation and are mixed with T.E. gas in a wall-less proportional counter. Ellett and Braby (1972) determined lineal energy spectra produced in spheres of simulated site diameters ranging from 0.5 to 5 µm in regions in which tritium was uniformly distributed. The yf(y) vs log y spectrum[16] in a 1 µm sphere peaked near 4.5 keV/µm as compared with about 1.5 keV/µm for 200 kVp x rays and 0.22 KeV/µm for ^{60}Co γ rays. For all of the radiations a maximum value of y≈15 keV/µm was observed.

When electrons of sufficiently high energy penetrate the enclosure of a wall-less counter they are accompanied by energetic delta rays. These are far more important than delta rays produced by ions because their LET is comparable to that of the primary electrons and (by convention) their energy may be up to half as high. An indication of the complexity of delta ray build-up and of the concomitant

---

[16] The data were given for the earlier quantity Y defined as ε/d, where d is the sphere diameter. Then y=1.5Y. Hence the figures given by Ellet and Braby have been multiplied by 1.5.

change of energy of the primary electrons appears in data by Braby and Roesch (1980) who measured lineal energy spectra of electrons in the range of energies from 0.5 to 2 MeV after passage through plastic absorbers of thicknesses between 0 and 0.36 g/cm². The simulated diameters were about 0.5, 1 and 2 μm. In these experiments the beam from an accelerator was collimated or scattered and entered a tank that was very large (80 cm diameter and 120 cm length) to minimize wall effects. Although y_D generally decreased with absorber thickness this was not always the case and there was a frequent but not entirely consistent decrease with increasing site diameter.

Most of the studies of electron microsimetry were made in connection with *radiotherapy* and involved energies in excess of 5 MeV. The LET of electrons in water differs by less than 10% from 0.2 keV/μm in the energy range from 1 to several hundred MeV. Because of the substantial delta ray ranges the mean lineal energy of the primaries and their associated delta rays is less than 0.2 keV/μm. However delta rays injected from electron traversals outside of this site can contribute larger values of y to the spectrum. This may be the reason for the bi-phasic spectrum from 18 MeV electrons obtained by Amols and Kliauga (1985) Fig.IV.34. Comparison of data obtained with a walled counter showed that y_D and

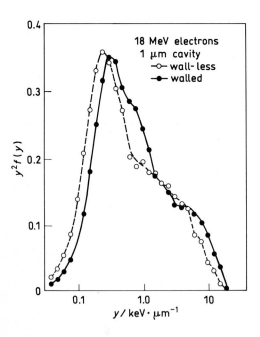

Fig.IV.34 Dose distributions, yd(y), due to 18-MeV electrons determined with a walled (solid line) and wall-less (dashed line) counter. Site diameter 1 μm (after Amols and Kliauga, 1985).

y_F were about 30% larger than in the wall-less counter. While the ratio of the diameters of the enclosure and the wall-less counter was 16:1 there is no certain assurance that even this was sufficient in view of the small angle with which delta rays are ejected by relativistic electrons. It is thus not clear whether the larger pulses are due to delta rays from primary electrons missing the counter or whether they may be due to a residual wall effect in which there is an excess of delta rays from the same primary electron.

Lindborg (1976) gave theoretical reasons why a wall-effect similar to that observed by Amols and Kliauga should obtain at 10 MeV. His measurements extended up to 39 MeV and he also derived appreciable corrections due to the *polarization effect* (Sect. IV.3.3) at 8 MeV. It should be noted that these effects operate in the same direction, i.e. a higher lineal energy in the gas than in the wall.

It must be concluded that the experimental microdosimetry of electrons is probably in a more unsatisfactory state than for other radiations. Better information is desirable in view of radiobiological data (Lloyd et al., 1975) indicating that the RBE of 15 MeV electrons relative to ^{60}Co γ radiations is 0.5 or less. It is of evident interest whether this can be explained on the basis of microdosimetric data.

IV.4.5 Ions

A.M. Kellerer has compared the track of an ion to a much-worn test-tube brush in which the central wire represents what is termed the core of the track and the bristles of varying length represent the delta rays produced in collisions between the ion and the atoms of the material in which the track is formed. To make the picture complete individual bristles are also frayed because of the production of secondary and higher order delta rays that arise from the primary ones. If the brush rotates and oscillates axially so that its image is blurred, the core becomes in essence a continuous uniform line which is surrounded by a *halo* (often somewhat improperly termed the *penumbra*) of cylindrical symmetry which corresponds to the average energy absorbed in the medium. This structure represents what Kellerer termed the amorphous track. The absorbed dose decreases approximately as the inverse square of the radial distance (Chatterjee and Schaefer, 1976) and a number of investigators have provided more accurate experimental data (Wingate and Baum, 1976; Mills and Rossi, 1980; Varma and Baum, 1980). There is evidently no sharp demarcation between core and halo. About 50% of the energy is deposited within a nanometer of the trajectory of ions having an energy of the order of MeV/amu. However even some of this fraction is deposited by low energy delta radiation. It has been estimated that about one half of the energy is deposited in energy transfer points located in the trajectory of such ions (Chatterjee and Schaeffer, 1976).

Ions produced by accelerators as well as alpha particles are generally monoenergetic and measurements with wall-less counters result in a triangular lineal energy spectrum that is modified by straggling and delta ray escape; this is accompanied by numerous small pulses that are due to delta rays injected into the counter by ions passing in its vicinity (Fig.IV.6). These make an important contribution to the f(y) spectrum but in the case of α particles of energies of a few MeV and site diameters of the order of 1 μm they are relatively unimportant when this spectrum is multiplied by y (and normalized) to derive the d(y) spectrum.

However, at high energies and in small sites the lineal energy is largely due to single δ rays generated outside of the site. In line with his general theoretical treatment of single and associated events in microdosimetry (section V.4.2) Kellerer has shown that y_D is due to two components. One of these is due to y_a, the expectation value of the amorphous track, and the other, y_δ, is due to single electrons. For ions of equal velocity y_δ is the same and y_a is proportional to L. This leads to the surprising result that a plot of y_D as a function of impact parameter (the distance between the center of the site and the ion trajectory) reaches a minimum just beyond the radius of the site where the track core is outside and y_D decreases sharply and there is also a sharp increase of y_δ. These characteristics are due to the fact that within the site the mean energy of individual delta rays is low compared to the energy deposited by the core while with the core outside the site only the more energetic delta rays reaching the site contribute to y_δ. This is illustrated in Fig.IV.35 (Kellerer and Chmelevsky, 1975)

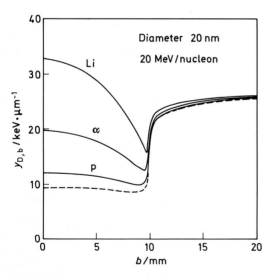

Fig.IV.35 Dose mean lineal energy $y_{D,b}$ in spherical sites at distances b from the trajectories of protons, alpha particles and lithium ions. The calculated values for 20-nm sites show a minimum when trajectories are tangent to the site (after Kellerer and Chmelevsky, 1975).

which shows computed y_D, the dose mean lineal energy average in 20 nm sites for Li ions, alpha particles and protons having an energy of 20 MeV/amu as a function of the impact parameter, b (the shortest distance between the ion trajectory and the center of the site). The distribution y_δ which is common to all of these ions is also shown.

An early observation which verifies the theoretical deduction is shown in Fig.IV.36 (Gross and Rodgers, 1972). The distributions for 4 MeV and 1.7 MeV protons in 0.25 μm sites exhibit larger y_a and smaller y_δ values for the lower energy where the LET is higher but the maximum delta ray energy is less.

Two studies of the microdosimetry of ions of energies in excess of 1 GeV have been reported. Lineal energy distributions from 3.9 GeV nitrogen ions have been determined at various depth (including some beyond the Bragg peak). A comparison between the spectra obtained with a walled and a wall-less counter (Fig.IV.37) shows the expected difference. The contribution by injected delta rays is absent in the walled counter in which back-scattered δ rays add to ion pulses (Sect. IV.3.5).

The spectra of delta rays from ions of even higher velocity (600 MeV/amu) were investigated by Metting et al. (1988) as function of the impact parameter employing a wall-less counter in a large enclosure (Sect. IV.3.7). Measurements at

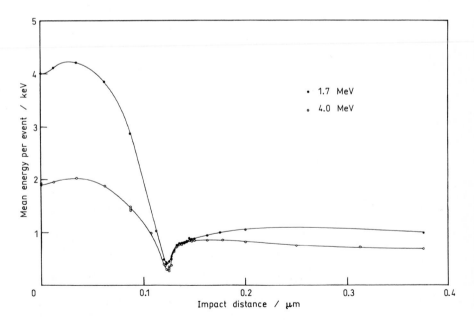

Fig.IV.36 Measured values of y_D as a function of distance from 1.7 and 4-MeV proton beams (after Gross and Rodgers, 1972). The scale of ordinate is arbitrary.

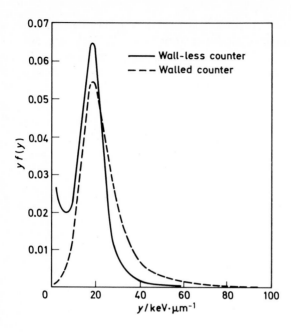

Fig.IV.37 Dose distribution d(y) of lineal energy due to 3.9 GeV nitrogen ions as determined with a wall-less (solid line) and a walled (dotted line) counter (after Rodgers et al, 1973).

this facility are described in section VI.2.2d. The site diameter was 1.3 μm and the measurements extended to impact parameters of 20 μm. Multiple delta rays contribute to the lineal energy spectrum up to about 5 μm. In agreement with other observations the spectrum of the single delta rays beyond this distance remained essentially constant. At b≈20 μm the average specific energy in events is more than 100 times larger than the absorbed dose as derived from the amorphous track with a corresponding event frequency of less than 10^{-2} per ion trajectory.

In view of the favorable depth dose distribution ions have been employed in radiotherapy. Because of the sharpness of the Bragg peak it is necessary to modulate the beam (usually with a motor driven disk incorporating absorbers of various thicknesses) in order to deliver a substantially uniform absorbed dose to the tumor volume (*spread Bragg peak*). It has been found that the RBE of energetic protons (~100 MeV) is higher than that of photons of comparable energy although their LET is nearly the same (i.e. essentially the minimum value for unit charge relativistic particles). It was suspected that the difference is due to the presence of lower energy ions created in nuclear spallations which was confirmed by microdosimetric measurements (Kliauga et al., 1978).

Microdosimetric spectra have also been obtained for 650 MeV helium ions in a therapy installation (Nguyen et al., 1976). The results confirm that in order to obtain constant biological effectiveness the absorbed dose in the spread Bragg peak should decrease with depth because of an increase of average lineal energy (and therefore of RBE).

IV.4.6 Pions

Negative pions (π^-) are heavy charged particles (rest mass 139.6 MeV) and thus display a Bragg-type distribution of dose as they penetrate tissue, and therefore better dose-localization properties then conventional (electrons, γ rays) radiation used in cancer therapy. Unlike other heavy charged particles, however, as they reach energies of the order of eV at the end of their range, negative pions are captured with high probability in atomic orbits forming what is termed a "pionic atom", that is an atom having a pion (rather than an electron) orbiting around the nucleus. Accompanied by Auger electron or x ray emission the orbital pion lowers its energy until it reaches the ground state of the atom. Because its mass is 273 times larger than the electron mass the pion will spend the majority of time inside the nucleus (more precisely, in the ground state there is very strong overlap between the negative pion's wave function and that of the nucleons). Being a strongly interacting particle the π^- is absorbed by the nucleus and its rest mass of 140 MeV re-appears in the form of kinetic energy of nuclear fragments: p, d, t, ^{3}He, ^{4}He, Li and neutrons, as well as γ rays. About 40 MeV of this is expanded to overcome the binding energy of the nucleons. This process, termed "star formation" because of the characteristic appearance of the pion tracks in nuclear emulsions, occurs very fast - in some 10^{-10} sec - and thus prevails over the usual decay mode of the negative pion ($\pi \to \mu + \nu$) that takes 100 time longer (2.54 10^{-8} sec). Pion capture generates not only additional dose in the Bragg peak but significant additional "effective" dose (i.e. dose multiplied by RBE) because most charged fragments emitted in a "star" are short ranged and have high LET. For similar reasons it is expected that the oxygen enhancement ratio (OER) of stopping negative pions will be lower than with low-LET radiation. These were, therefore, the reasons for proposing negative pions for radiotherapy, at least based on their physics and the known radiobiology of other heavy charged particles.

Microdosimetric distributions at various depth have been measured by a number of investigators. A study employing the time-of-flight technique and fast scintillators has permitted separate assessment of the lineal energy spectra of passing as well as stopping pions and also those of contaminating electrons and muons (Schuhmacher et al., 1985). The authors found a simple relation between the total absorbed dose, D, and the fraction D^* due to stars and y_D of the total spectrum

$$\frac{D^*}{D}(\%) = \frac{0.99 y_D}{keV \ \mu m^{-1}} - 3.3 \qquad (IV.23)$$

The maximal deviation from this formula was no more than 2.5%. This permits determination of D^* with routine microdosimetry.

Further information on measurements at a pion therapy facility is presented in section VI.2.2.

IV.5 Measurement of Distributions of Specific Energy

IV.5.1 General Comments

Unlike lineal energy, y, which is a measure of the energy imparted to a site in an individual event, the mean specific energy, z, is proportional to the total energy imparted during an irradiation that is usually quantified in terms of the absorbed dose, D. In general, the specific energy is imparted in a number of events that depends on D, LET and site diameter, d. It varies if these quantities have the same values according to the distribution f(z). f(y) is a distribution that is suitable to characterize the quality of radiation, while f(z) represents the spectrum of energies deposited in sites and therefore can serve to generally characterize the effects of radiation if these depend on the energy deposited in sites regardless of its temporal and geometrical distribution. The effect probability is then

$$E = 1 - e^{-kD} \qquad (IV.24)$$

where

$$k = \int_0^\infty e(z)f(z)dz \qquad (IV.25)$$

with the effect probability e(z) for specific energy z.

An event of lineal energy, y, produces an increment of z that is usually expressed in a relation that applies to a unit density spherical site and employs the non-coherent units of Gy for z, keV/µm for y and µm for d:

$$z \ / \ Gy = 0.204 \frac{y \ / \ \mathrm{KeV} \ \mu m^{-1}}{d^2 \ / \ \mu m^2} \qquad (IV.26)$$

$f_1(z)$ the distribution of specific energies in single events therefore differs from f(y) only by a constant factor. Since events are, by definition, statistically independent, it is possible to calculate f(z) for any absorbed dose according to a procedure given in Chapter II.

Early measurements of f(z) employed a technique which involved a simple modification of the operation of the differential analyzer that records f(y) spectra (Rossi et al, 1961). In this device analog to digital conversion results in a pulse train that is proportional to y. Instead of registering the number of pulses resulting from single events, those from various numbers of events were added during a time period corresponding to an absorbed dose $\sigma^2 D$ (where σ, the density of the tissue equivalent gas in the counter relative to that of tissue, is of the order of 10^{10}). In the automated procedure 1000 or more measurements of z could be made and f(z) even for large absorbed doses from weak sources could be determined rapidly. Figure IV.38 shows f(z) in 7 µm diameter sites irradiated by ^{60}Co gamma radiation to absorbed doses ranging from 0.75 mGy to 7.5 Gy.

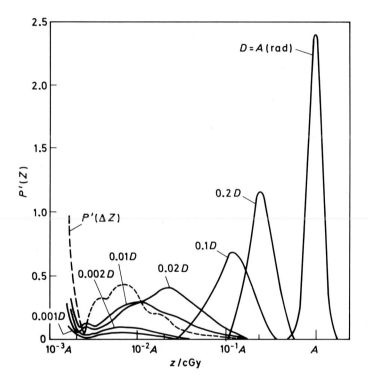

Fig.IV.38 Frequency distributions zf(z) of specific energy due to ^{60}Co radiation. Absorbed doses between 0.75 mGy and 7.5 Gy (after Rossi et al, 1961). The scale of ordinate is arbitrary. The dotted line is the f_1 (z) distribution.

IV.5.2 The Variance Method

After the early measurements of f(z) interest in experimental determinations of f(z) declined because of the ready availability of computers and programs permitting calculations of this distribution based on the knowledge of f(y). Furthermore, the utility of f(z) appeared to be limited to certain phases of theoretical radiobiology. However, its experimental determination regained importance because of two reasons. One of these is its relevance to radiation protection and the other is its potential value in structural microdosimetry.

Bengtsson (1969) drew attention to the importance of the fact that a knowledge of f(z) permits the determination of y_D, the dose average lineal energy. In a good first approximation this quantity is proportional to the quality factor, Q, in radiation protection when the LET is less than about 200 keV/μm (Sect.VI.3) and, for essentially the same reason, the related quantity, z_D (the dose average of specific energy from single events) is of importance in radiobiology and specifically in the *theory of dual radiation action* (Sect.VI.1.4).

Measurements of f(z) can often be performed with ionization chambers rather than the more complex proportional counters because a substantial number of events can be accepted in each measuring cycle. This also permits measurements at pulsed radiation sources where pulse pile-up in proportional counters often makes a measurement of f(y) impossible because of unacceptably low counting rates.

The relation that is basic to the *variance technique* is that the quantity z_D which is proportional to y_D is given by:

$$z_D = \frac{\overline{z^2}}{\overline{z}} - \overline{z} \tag{IV.27}$$

with

$$\overline{z^2} = \frac{1}{I} \sum_{i=1}^{I} z^2{}_i , \qquad \overline{z} = \frac{1}{I} \sum_{i=1}^{I} z_i . \tag{IV.28}$$

These quantities are determined as the averages of a large number, I, of individual measurements (i) that are automated. Another method introduced by Bengtsson (1972) employs a circuit that measures the fluctuations of the current in an ionization chamber.

Lindborg (1974) utilized this method to determine z_D for low LET sources that had the constant radiation output necessary because fluctuations in intensity would cause additional broadening of the spectrum with erroneously high indicated z_D values. It was however recognized that this source of error could be eliminated if each value of z_i is divided by a quantity that is proportional to the intensity during the measurement interval (as for instance the value of the beam current of an ion accelerator when z_D of the neutrons from its target is determined).

Kellerer and Rossi (1984) showed that the second instrument may in fact determine the correction factor with a variance of its own. In this method a measurement is simultaneously made with a counter and another detector which can also be a counter that may, but does not need to, have identical characteristics. This has become known as the *variance-covariance method* although the adjective *relative* applies to both terms.

The relative variance of specific energy in the counter is

$$V_a = \frac{\overline{z_a^2}}{\overline{z}_a^2} - 1 \tag{IV.29}$$

and the relative variance of the signal, s, in the second detector is

$$V_s = \frac{\overline{s^2}}{\overline{s}^2} - 1 \tag{IV.30}$$

If the second detector is also a proportional counter s may be replaced by z_s.

The relative covariance of the two devices is

$$C_{as} = \frac{\overline{z_a s}}{\overline{z}_a \overline{s}} - 1 \tag{IV.31}$$

where the numerator of the fraction refers to the product of the concomitant values of these quantities as determined in a large number (typically 1000) of measurements. The dose average specific energy per event is then

$$z_D = (V_a - C_{as})z_a \tag{IV.32}$$

When the second instrument is also a proportional counter the corresponding expression is

$$z_D = (V_s - C_{as})z_s \tag{IV.33}$$

and any differences between these quantities are due to statistical limitations or other sources of error which can be reduced by accepting their average as the value of z_D.

It is essential that the two devices operate independently. Thus the signals from an energetic particle that traverses both counters are entirely covariant and would thereby lend to an underestimate of z_D. The kind of detector employed in such measurements is shown in Fig.IV.22.

Modern signal processing techniques permit ready implementation of this method and, because little or no gas amplification is required, it has been possible to determine z_D (and hence y_D) for very small sites. The increase, due to straggling, in y_D for neutrons (Goldhagen and Randers-Pehrson, 1992) and the even greater one for γ radiation (Chen et al., 1990) is shown in Figs.IV.39 and IV.40. Data of this kind can be employed in the experimental determination of the proximity function in structural microdosimetry (see Sect. V.4.3.1).

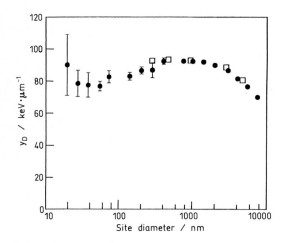

Fig.IV.39 Dose mean lineal energy, y_D, due to 15 MeV neutrons as a function of site diameter. Determinations employing standard pulse height measurements in a single counter (□) or based on variance covariance technique (●) (after Goldhagen and Randers-Pehrson, 1992).

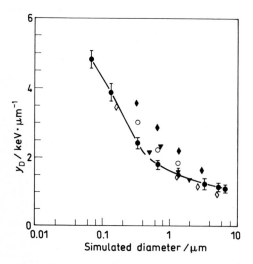

Fig.IV.40 Dose mean lineal energy, y_D, due to ^{137}Cs gamma radiation determined with the variance-covariance method (after Chen et al, 1990).

The use of variance measurements in radiation protection assumes that the quality factor, Q, is proportional to y_D. Although this may be an adequate approximation in many instances it can lead to substantial discrepancies at high values of y_D. Zaider and Rossi (1989) have shown that when $1/z_F$, the number of events per unit absorbed dose, is determined with an auxiliary device, the distribution f(y) can, in principle be computed. Statistical limitations usually make it impossible to derive more than the third moment of this distribution. However, information on even the first two moments should be sufficient in most radiation protection applications.

IV.6 Measurement of LET Distributions

The measurement of d(L), the normalized distribution of absorbed dose in LET (Rossi and Rosenzweig, 1955), which was the impetus to the development of microdosimetry, is now infrequently practiced. Even with the quality factor, Q(L), based on LET, radiation protection measurements usually determine the dose equivalent by multiplication of D(y) by Q(y) which is an approximately equivalent quantity (see Sect.VI.3).

The measurement of d(L) is based on the "LET assumption", i.e. that the absorbed dose is imparted to the gas in the proportional counter by charged particle tracks that are straight lines in which there is constant and continuous energy loss. All tracks are assumed to traverse the gas volume, i.e. the particle range is very much longer than the counter diameter. The fact that there can be major divergence from these postulates is a basic reason for the existence of microdosimetry, but they are often sufficiently realistic to permit assessments of the d(L) spectrum with adequate accuracy. This applies in the most important practical application when the spectrum is due to neutrons of energies exceeding a few hundred keV and the site diameter is of the order of 1 μm or less.

On the basis of these assumptions, particles of fixed LET equal to L impart an energy that is equal to the product of L and the length of the chord formed by the track intercept. With the counter uniformly irradiated the fluence is constant over the cross-sectional area of the cavity, a condition termed μ *randomness* (Kellerer, 1981). In a sphere of diameter d the chord length distribution, f(x), is triangular:

$$f(x) = \frac{2x}{d^2} \qquad x \in [0, d] \tag{IV.34}$$

and the basic characteristics of the spectrum of energy depositions can be visualized by considering the joint contribution of a few groups of particles of discrete LET (Fig.IV.41). This makes it evident that complete information on the spectrum is obtained by registration of pulse heights down to a minimum, P_{min},

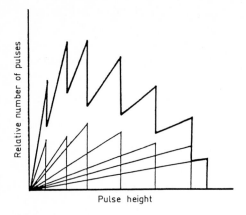

Fig.IV.41 Superposition of triangular spectra resulting from particles of discrete LET values traversing a spherical cavity when straggling, scattering etc. are neglected (after Rossi and Rosenzweig, 1955).

corresponding to a traversal along a major diameter by particles of least LET because below this value the spectrum decreases linearly to zero. Hence a proper analysis can include the contributions of pulses lost when the discriminator required to suppress noise is set anywhere below P_{min}.

A simple integration (Rossi, 1968) leads to the result that D(L) the distribution of absorbed dose in LET is given by:

$$D(L) = Dd(L) = \frac{2}{\pi d^2}\left[N(E) - E\frac{dN(E)}{dE} \right] \qquad (IV.35)$$

where N(E) is the number of energy depositions by individual particles per unit interval of E. The value of E corresponding to a pulse height, P, (e.g. P as determined by the differential analyzer) is $\eta P/P_0$ where P_0 is the pulse height when the counter receives an energy η from a calibrating source. As indicated in Eq. IV.35 D(L) is not normalized but represents the distribution of the absorbed dose, D, in the counter volume or in the simulated region of unit density tissue.

The pulse height spectra obtained in practice are usually highly skewed and are commonly plotted in logarithmic representation. The slope, S, at E is then

$$S = \frac{d \log N(E)}{d \log E} = \frac{E}{N(E)}\frac{dN(E)}{dE} \qquad (IV.36)$$

which, substituted in Eq IV.35 yields

$$D(L) = \frac{2}{\pi d^2} N(E) (1 - S) \qquad (IV.37)$$

This permits a simple graphic evaluation of D(L). A numerical technique for the evaluation of D(L) has been provided by Kellerer (1972).

Because of the limited validity of the underlying assumptions, the distribution, D(L) as determined by this technique depends on d and for any value of d one obtains only an approximation to the abstract spectrum that can only be determined by calculation. This is illustrated in Fig.IV.42 which shows the theoretical spectrum[17] for 2 MeV neutrons as derived by Boag (1954) and experimentally determined spectra for d=0.75 μm and d=3 μm. The theoretical spectrum which has a singularity at an LET corresponding to the Bragg peak is distorted by straggling which is less for d=3 μm at low LET but the distribution does then not extend to the theoretical maximum value of L because the maximum average LET of a proton traversing 3 μm of unit density tissue is near 75 keV/μm. On the other hand the theoretical maximum is exceeded in small distances because of straggling.

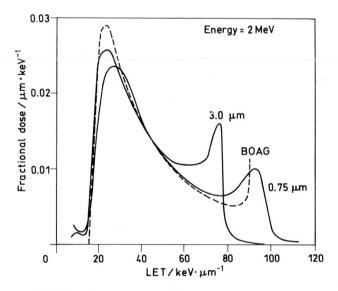

Fig.IV.42 Distribution of absorbed dose in LET due to 2-MeV neutrons. Measurements in 0.75 and 3 μm sites and a theoretical curve by Boag (1954).

---

[17] The calculation is limited to the very dominant contribution of recoil protons.

IV.7 Appendix

The V Effect

The simplest type of nuclear spallation consists in the form of a V-track, i.e. a pattern in which a pair of charged nuclear fragments is projected from a common vertex at a relative angle 2θ. Spallations generally occur at high energies where 2θ is likely to be small because of relativistic effects, and the particle range is likely to be large. Under these conditions avoidance of the wall effect (section IV.3.5) may require wall-less counters to be placed into very large enclosures to avoid coincidences of the particles when they should be registered separately.

This analysis of the V-effect is also suitable in cases where spallation results in more than two charged fragments, because it applies to any pair of them. The analysis also applies to other instances where a pair of charged particles is propagated from a vertex in rectilinear trajectories, e.g. those of a scattered and of a recoiling heavy charged particle. Because the trajectories of electrons, and especially of delta rays, are generally *not* rectilinear the results of this analysis are only of limited applicability to them.

Figure IV.43 shows a spherical wall-less counter of radius R inside a spherical enclosure of radius K. It is only necessary to consider pairs having a bisector, B, that is orthogonal to the surface of the enclosure wall because when the angle is different the geometry is the same as that for pairs originating at another point on the surface where the bisector is orthogonal, i.e. the situation applies whether bisectors are unidirectional or isotropic.

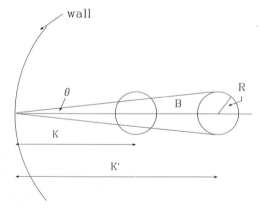

Fig.IV.43 A wall-less counter of radius R is enclosed in a spherical cavity delimited by a solid wall. K is the distance between the center of the counter and the wall, as indicated. A pair of particles originate from the wall with relative angular separation, 2θ. When K>K' only one the particles may intersect the counter.

If the vertex of the pair is inside the enclosure the objective of the wall-less counter is met because there is no change of density involved (see section IV.3.5). If the distance between the center of the counter and the wall is equal to, or greater than K'=R/sinθ this is evidently also the case and no coincidences can occur even when the vertex is at the wall surface.

The same relative relations exist within the wall material but the absolute values are greatly reduced by ρ, the relative density of the gas with respect to this material. Thus when the vertex is at a depth exceeding K′ρ the separation of the fragments is large enough so that there can be no coincidence originating from the wall. A typical value of ρ is 10^{-5} and in the example of Fig.IV.43 K' is about 5 times larger than the counter diameter. Hence if the simulated site diameter is 1 μm there should be no coincidence when the vertex is at a depth of about 5 μm and the emergent particles are separated by 1 μm. This is virtually equivalent to a point source at the wall surface and if the (physical) diameter of the counter is more than 1 cm this geometry is closely approached even for depths as large as 1000 ρK' where the separation of the emerging particles is 1 mm. At very high energies when the range of the particles in the wall material exceeds several centimeters this may not be the case and in this unusual situation, and only for the portions of the track that are some centimeters from the vertex, the counter-to-wall distance may be less than K'.

It may be thus concluded that the requirement that K>K' insures that regardless of particle range, the relative frequency and the nature of coincidences in the counter are the same as those in the wall material for θ or any larger half angle. However, this criterion (K>K') may not be met in practice and it is of interest to determine the error involved if K is smaller.

Figure IV.44 shows the geometry of the problem. The simplification is introduced that the wall surface is flat rather than spherical. In most cases (e.g. K>4R) this introduces a negligible error and it may be assumed that (in the usual situation of a uniform radiation field) the relative number of pairs originating in the interval db about b is 2πbdb. A counter of radius R has the center at A(0,b,K) and pairs of charged particle tracks originate at O with azimuthal angles φ and π+φ, respectively and inclination θ relative to the z axis. b is the lateral distance of the center of the counter (measured from the z axis) and K is the distance from the wall (represented by the plane x,y) to the equatorial plane of the counter that is parallel to the wall (see Fig.IV.44). We shall focus for the moment on the question of finding, for one single track, the combined ranges of b and φ for which the track intersects the counter. The other parameters of the problem (R, K and θ) will be considered fixed and given.

The equation of the sphere is:

$$x^2 + (y - b)^2 + (z - K)^2 = R^2 \qquad (IV.38)$$

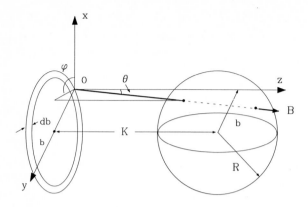

Fig.IV.44 A track segment (OB) intersects a spherical counter of radius R with the center at (0,b,K). The second track segment, with azimuthal angle $\phi + \pi$, is not shown. The question asked is to determine the range of values b and $\phi(b)$ for which OB intersects the sphere; and at each b to find those ϕ values for which *both* track segments intersect the counter.

and the equation representing the track is:

$$x = t \sin \theta \cos \phi$$
$$y = t \sin \theta \sin \phi \qquad\qquad\qquad\text{(IV.39)}$$
$$z = t \cos \theta$$

where t is a parameter. To obtain the coordinates of the point(s) where the track OB intersects the sphere, x,y,z of Eq(IV.39) are inserted in Eq(IV.38). One obtains a quadratic equation for t:

$$t^2 - 2(b \sin \theta \sin \phi + K \cos \theta)\, t + (b^2 + K^2 - R^2) = 0 \qquad \text{(IV.40)}$$

To have intersection points Eq(IV.40) needs to have real solutions, $t_{1,2}$. The usual condition for this is:

$$\Delta = (b \sin \theta \sin \phi + K \cos \theta)^2 - (b^2 + K^2 - R^2) \geq 0 \qquad \text{(IV.41)}$$

Before proceeding further note the following:
a) because of the symmetry of the problem only positive values of b need to be considered,
b) $\theta \in [0, \pi/2]$ so $\sin\theta$, $\cos\theta$ and $\tan\theta$ are all non-negative, and
c) one is concerned here only with the case K>R (i.e. tracks originate from outside the counter). The condition, Eq(IV.41), is satisfied either if:

$$1 \geq \sin \phi \geq \frac{\sqrt{b^2 + K^2 - R^2}}{b \sin \theta} - \frac{K}{b \tan \theta}, \qquad \text{(IV.42)}$$

or if:

$$-1 \leq \sin \phi \leq -\frac{\sqrt{b^2 + K^2 - R^2}}{b \sin \theta} - \frac{K}{b \tan \theta}. \qquad \text{(IV.43)}$$

For the inequalities, Eq(IV.42), to be possible the upper limit of $\sin\phi$ must be smaller than 1:

$$\frac{\sqrt{b^2 + K^2 - R^2}}{b \sin \theta} - \frac{K}{b \tan \theta} \leq 1 \qquad \text{(IV.44)}$$

which is in fact a condition for b; it obtains if:

$$0 \leq b \leq K \tan \theta + \frac{R}{\cos \theta}. \qquad \text{(IV.45)}$$

As for the other inequalities, Eq(IV.43), it may be readily shown that if, as assumed, b>0 and K>R then they are not satisfied for any value of ϕ. To summarize the results obtained so far: the track segment intersects the counter only if b satisfies Eq(IV.45); for each value of b in this range the azimuthal angle, ϕ, must be in the interval:

$$\phi \in [\phi_0, \pi - \phi_0] \qquad \text{(IV.46)}$$

where, by definition:

$$\phi_0(b) = Max \left\{ -\frac{\pi}{2}, \arcsin \left[\frac{\sqrt{b^2 + K^2 - R^2}}{b \sin \theta} - \frac{K}{b \tan \theta} \right] \right\} \qquad \text{(IV.47)}$$

We consider next the question of two track segments (whose azimuthal angles differ by π) intersecting the spherical counter.

a) If ϕ_0 is in the interval $[0, \pi/2]$ then the two segments can not intersect the sphere simultaneously (see Fig.IV.45). From the definition of ϕ_0, Eq(IV.47), this happens when:

$$b \geq \sqrt{R^2 - K^2 \sin^2 \theta}. \qquad \text{(IV.48)}$$

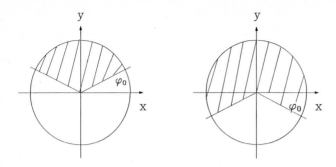

Fig.IV.45 The dashed region represents the range of ϕ values where at least one track segment intersects the counter. Only in the figure at the right (where ϕ_0 is negative) two segments with azimuthal angles differing by π can intersect the counter simultaneously.

b) Otherwise, when

$$b \in \left[0, \sqrt{R^2 - K^2 \sin^2 \theta}\,\right] \tag{IV.49}$$

both segments will intersect the sphere whenever the angle ϕ is restricted to (see Fig.IV.45):

$$\phi \in \left[\phi_0, -\phi_0\right] \cup \left[\pi + \phi_0, \pi - \phi_0\right]. \tag{IV.50}$$

Let P represent the probability that a track segment that intersects the counter does so in coincidence with the other segment. At a *given* b, the range of ϕ in Eq(IV.50) is $4|\phi_0(b)|$; and the *total* range of ϕ, so that at least one of the two segments will intersect the sphere, is $\pi - 2\phi_0(b)$ [see Eq(IV.46)]. As already indicated, the number of track segments originating in the interval [b, b+db] is proportional to $2\pi b db$. From this, one obtains the result:

$$P = \frac{\displaystyle\int_0^{\sqrt{R^2 - K^2 \sin^2 \theta}} (2\pi b)\,4|\phi_0(b)|\,db}{\displaystyle\int_0^{K \tan \theta + R/\cos \theta} (2\pi b)\left[\pi - 2\phi_0(b)\right]db}. \tag{IV.51}$$

A plot of P as a function of θ is shown in Fig.IV.46.

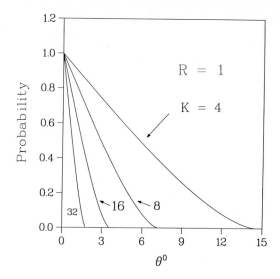

Fig.IV.46 The probability that a track intersects the counter together with its companion spallation track as a function of the angle θ. The two tracks are at relative angle 2 θ(see text).

Chapter V
Theoretical Microdosimetry

V.1 A Diversion in Geometric Probability

In Chapter II it was shown that the specific energy spectrum can be factorized in terms of a dose-dependent term and a z-dependent one. In *regional* microdosimetry a second factor that contributes to the shape of the microdosimetric distribution is the geometry of the site. The question then arises as to whether it would be possible to further factor out this contribution. This possibility is suggested by the fact that - in a homogeneous medium - one can regard the spectrum of energy deposition as resulting from the random overlap of two "objects": the track (i.e. a collection of transfer points) and the site. As a matter of fact, this is precisely the way most theoretical microdosimetric distributions are generated, namely by randomly placing a site on a Monte-Carlo-generated track (see Section V.2). The interesting (and powerful) consequences of this viewpoint are examined in the following through a brief excursion in the domain of geometric probability. For a general introduction to this subject the book by Santalo (1976) may be consulted.

Consider two objects (3-dimensional domains) of volume V_1 and V_2 (we shall label an object with its measure, here volume; this should create no confusion). To each object one may assign a fixed point (a "center") and a fixed vector which determines its orientation. We assume that these objects are randomly placed with their centers in a domain S according to some (unspecified) probability distribution (Fig.V.1). Let v be the measure (volume) of the overlap of V_1 and V_2 [$v=V_1 \cap V_2$] and let further $\rho(v)dv$ be the probability measure of v, that is

$$\Pr[v \in V_0] = \int_S C(v; V_0)d\rho(v) \qquad (V.1)$$

where for any domain V_0,

$$C(v; V_0) = \begin{cases} 1 \text{ if } v \in V_0 \\ 0 \text{ otherwise} \end{cases} \qquad (V.2)$$

For every point $r \varepsilon S$ and volume v define

$$g(r, v) = \begin{cases} 1 \text{ if } r \in v \\ 0 \text{ otherwise} \end{cases} \qquad (V.3)$$

S

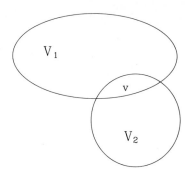

Fig.V.1 Two objects of volume V_1 and V_2 are placed randomly with their centers (not shown) in the domain S. Their intersection has volume v. Only the case v'0 is considered in the text.

Clearly

$$v = \int_S g(\mathbf{r}, v) \, d^3 r \qquad (V.4)$$

We are interested in the expected (average) value of v, denoted E(v). This can be obtained as follows :

$$E(v) = \int v d\rho(v) = \int\int g(\mathbf{r}, v) \, d^3 r d\rho(v) = \int_S \Pr[\mathbf{r} \in v] d^3 r \qquad (V.5)$$

A similar result can be obtained for $E(v^2)$:

$$E(v^2) = \int v^2 \, d\rho(v) = \int\int\int g(\mathbf{r}_1, v) g(\mathbf{r}_2, v) \, d^3 r_1 \, d^3 r_2 \, d\rho(v) =$$
$$= \int\int \Pr[\mathbf{r}_1 \in v, \ \mathbf{r}_2 \in v] d^3 r_1 \, d^3 r_2 \qquad (V.6)$$

The general result is

$$E(v^k) = \int \ldots \int \Pr[\mathbf{r}_1 \in v \ and \ \mathbf{r}_2 \in v \ and \ldots r_k \in v] d^3 r_1 \, d^3 r_2 \ldots d^3 r_k$$
$$(V.7)$$

The expression, Eq(V.7), has been obtained by Robbins (1944); it shows how to calculate the expected values of any of the moments of v if we know the probability that a point belongs to v. Note that

$$\Pr(\mathbf{r}_1 \in v \ and \ \mathbf{r}_2 \in v \ and \ldots and \ \mathbf{r}_k \in v) \neq [\Pr(\mathbf{r} \in v)]^k \qquad (V.8)$$

since the volume v is not fixed but is subject to a probability measure $\rho(v)$.

To make further progress we shall assume that V_1 and V_2 are uniformly and isotropically distributed. Consider now the following particular situation defining v: (a) V_2 consists of a spherical shell of radius r and thickness dr; then $V_2 = 4\pi r^2 dr$, and (b) the center, r_A, of V_2 is always inside V_1 (Fig.V.2). Let

$$U(r) = \frac{E(v)}{4\pi \, r^2 \, dr} \qquad (V.9)$$

be the expected fraction of V_2 overlapping V_1. From Eq(V.5):

$$E(v) = \int_{V_1} \Pr(r_0 \in v) \, d^3 r_0 = \int_{V_1} \Pr(r \in V_1 | r_A \in d\,r_A) \, d^3 r_A$$
$$= 4\pi \, r^2 \int_{V_1} \Pr(r \in V_1 | r_A \in d^3 r_A) \, d^3 r_A \qquad (V.10)$$

The notation $\Pr(X|Y)$ means the conditional probability of X given Y. From Eqs(V.9,10):

$$U(r) = \Pr(r \in V_1 | r_A \in V_1) \qquad (V.11)$$

U(r) represents then the probability that - given a random point in V_1 - a second point isotropically placed at distance r will also belong to V_1. U(r) is obviously a property of the domain V_1. A quantity playing a similar role is the *point-pair distribution of distances*, p(r); it is defined as the distribution of distances between pairs of points randomly chosen in the domain. Thus

$$p(r)dr = \frac{E(v)}{V_1} = \frac{U(r)4\pi r^2 dr}{V_1} \qquad (V.12)$$

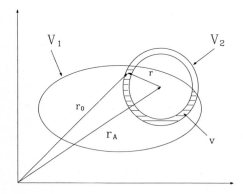

Fig.V.2 The geometrical arrangement used in Eqs(V.9-12).

As shown by Kellerer (1986) it is possible to obtain a relation between the expected value of v^2 and the proximity functions of the two overlapping domains, V_1 and V_2. Indeed, from Robbins' result, Eq(V.6), we have

$$E(v^2) = \int\int \Pr[r_1 \in v \text{ and } r_2 \in v] \, d^3 r_1 \, d^3 r_2 =$$
$$= \int \Pr(r_2 \in v)\left[\int \Pr(r_1 \in v | r_2 \in v) \, d^3 r_1\right] d^3 r_2 \qquad\text{(V.13)}$$

Since V_1 and V_2 are isotropically distributed in space the conditional probability in the second term of the expression above should depend only on $|r_2-r_1|=r$. Furthermore,

$$\Pr(r_1 \in v | r_2 \in v) = U_1(r)\,U_2(r) \qquad\text{(V.14)}$$

where U_1 and U_2 correspond to V_1 and V_2, respectively. Indeed, given that $r_2 \in v$, we have $r \in v$ only if $r \in V_1$ *and* $r \in V_2$; since these two conditions are met *independently*, Eq(V.14) follows. With this

$$E(v^2) = \int \Pr(r_2 \in v)\left[\int U_1(r)\,U_2(r)4\pi\, r^2 dr\right] d^3 r_2 \qquad\text{(V.15)}$$

However, from Eq(V.5)

$$\int \Pr(r_2 \in v)d^3 r_2 = E(v) \qquad\text{(V.16)}$$

and therefore

$$\frac{E(v^2)}{E(v)} = \int U_1(r)\,U_2(r)4\pi r^2 dr \qquad\text{(V.17)}$$

or, equivalently [see Eq(V.12)]:

$$\frac{E(v^2)}{E(v)} = V_1 V_2 \int \frac{P_1(r)\,P_2(r)}{4\pi r^2} dr \qquad\text{(V.18)}$$

This important result can be immediately translated to microdosimetry. Assume for this purpose that instead of using spatial dimensions as a measure of volume we use the energy deposited. The "volume" occupied by each transfer point equals the energy locally deposited, v becomes the energy, e, deposited in the site, V_1 becomes the total energy in the track, T, and V_2 is the volume of the site:

$$\frac{E(e^2)}{E(e)} = T V_2 \int \frac{P_{track}(r)\,P_{site}(r)}{4\pi r^2} dr. \qquad\text{(V.19)}$$

The product between the point-pair distance distribution and the measure of the site is termed the *proximity function* of that domain. Thus we have a proximity function of energy deposition,

$$t(r) = (total\ energy)\ P_{track}(r) \qquad\qquad (V.20)$$

and also a proximity function of the site:

$$s(r) = (site\ volume)\ P_{site}(r) \qquad\qquad (V.21)$$

$t(r)dr$ can be also defined as the expected energy imparted to a shell of radius r and thickness dr centered at a randomly chosen transfer point. It is clear, however, that - since we assumed the "volume" of each transfer point to be measured by the energy locally deposited - that the random selection of each point is being made proportional to this local energy. The expression, Eq(V.19), represents the *dose-averaged energy deposited*. A similar expression can be written in terms of the dose-averaged specific energy,

$$z_D = \int \frac{t(r)\ s(r)}{4\pi r^2 m}\ dr \qquad\qquad (V.22)$$

where m is the mass contained in the site.

V.2 Monte Carlo Simulation of Charged-Particle Tracks

A standard method for calculating microdosimetric distributions, proximity functions, radial dose distributions and related quantities is to use computer-generated particle tracks. To the extent that accurate cross sections for the interaction of the ionizing particles with the medium are known this method represents the best procedure for obtaining microdosimetric quantities. The technique used for generating particle tracks is *Monte Carlo* sampling; a brief introduction to this technique is given below.

V.2.1 A Brief Visit to Monte Carlo Sampling

The Monte Carlo method consists of mathematical techniques for sampling values of a random variable, x, given its cumulative distribution function, F(x). The development of this method, in the early 1940's, is attributed to John von Neumann and Stanley Ulam who applied it - as remains largely the case today - to problems in physics and nuclear engineering. We shall use the following notations:

Random variables are designated by X,Y,... and their numerical realizations by x,y,..., respectively. For instance, if X represents the number of radioactive disintegrations per second from a certain source then x may assume the values 0,1,2,... The *cumulative distribution function* (cdf) of X, termed F(x), is the probability that the random variable X assumes a value equal or less than x:

$$F(x) = Prob \{X \leq x\} \qquad (V.23)$$

Given that F(x) is continuous over the interval [a,b], f(x) is termed the *probability density function* (pdf) of X

$$F(x) = \int_a^x f(x')d'x, \quad a \leq x \leq b. \qquad (V.24)$$

f(x)dx is often described as the probability that x≤X≤x+dx:

$$f(x)dx = Prob \{X \in [x, x + dx]\} \qquad (V.25)$$

However, the probability that X takes any particular value, x_0, is zero if F(x) is continuous. Clearly

$$f(x) = \frac{dF(x)}{dx}. \qquad (V.26)$$

It is convenient (but not necessary) to normalize F(x) to unity, that is, if the domain of X is [a,b] then

$$F(b) = \int_a^b f(x)dx = 1 \qquad (V.27)$$

We shall assume, in the following, the normalization, Eq(V.27).

The simplest (although not always possible) method of sampling x according to F(x) is the following: Consider the random variable Y the realizations of which, y, are obtained as:

$$y = \int_a^x f(x')d'x = F(x), \quad a \leq x \leq b \qquad (V.28)$$

Since F(x) is normalized, the domain of y is [0,1]. What is the cdf of y ? F(x) is a *monotonically* increasing function of x. Then

$$G(y) = Prob\{Y \leq y\} = Prob\{X \leq x\} = F(x) = y \qquad (V.29)$$

or

$$g(y) = \frac{dG(y)}{dy} = 1, \quad 0 \le y \le 1 \tag{V.30}$$

and we have proved that y is *uniformly* distributed in the interval [0,1]; generators of uniformly distributed numbers (hereafter denoted by ξ) in [0,1] are routinely available on modern computers. To sample x one proceeds as follows:

1. Obtain ξ
2. Calculate $x = F^{-1}(\xi)$, see Eq(V.29).

This method is applicable whenever $F^{-1}(x)$ exists. For instance, consider

$$f(x) = \alpha\, e^{-\alpha x}, \quad F(x) = 1 - e^{-\alpha x}, \quad \alpha > 0, \quad 0 \le x < \infty \tag{V.31}$$

Then

$$x = -\frac{1}{\alpha} \log(1 - \xi) \tag{V.32}$$

or, equivalently

$$x = -\frac{1}{\alpha} \log \xi' \tag{V.33}$$

since, if ξ is uniformly distributed in [0,1] so is $\xi' = 1 - \xi$.

The method just described illustrates the general procedure used in Monte Carlo simulation: through a series of transformations one attempts to reduce the problem to that of generating pseudorandom numbers, ξ.

Quite often, information on $F(x)$ comes from experiment and therefore we have a set of discrete values $y_i = F(x_i)$, i=1,2,...,n. If the analytical form of $F(x)$ is known, or obtainable by fitting a parametric expression to the data $\{y_i\}$, and if the inverse function $F^{-1}(x)$ exists one may still apply the procedure described above. Otherwise $F(x)$ is approximated with a histogram:

$$Prob\{y_i \le Y \le y_{i+1}\} = \frac{y_i}{y_T}, \tag{V.34}$$

where

$$y_T = \sum_{i=1}^{n} y_i. \tag{V.35}$$

Imagine on a straight line adjacent segments of length y_i/y_T (i=1,2,...,n). The total length of this construct is 1, and a random number, ξ, will necessarily be contained in one of the segments and thus select a value x_j. This is described in the following algorithm:

1. select a number ξ,
2. calculate sequentially

$$g_i = g_{i-1} - \frac{y_i}{y_T} \quad (g_0 = \xi, \ i = 1,2,\ldots,n) \tag{V.36}$$

and select the x_i corresponding to the first negative g_i.

This method is useful for any discrete random variable X, the realizations of which take a finite number of values, n.

The *rejection* technique applies to probability distribution functions, f(x), for which $a \leq x \leq b$ and

$$f^* = \max[f(x)] \tag{V.37}$$

is known and finite. f(x) is contained in a rectangle of length (b-a) and height f^*. The sampling algorithm is:

1. Select two numbers ξ_1 and ξ_2
2. Calculate

$$x = a + (b - a)\,\xi_1 \tag{V.38}$$

3. Accept x if $\xi_2 \leq f(x)/f^*$. If not, return to 1.

The rejection technique is illustrated in Fig.V.3. In terms of computing time the efficiency of this technique (i.e. the relative number of "successes" at step 3 above) is given by the ratio $1/[f^*(b-a)]$.

A very efficient algorithm applicable in most cases (Kahn, 1956; Messel and Crawford, 1970) takes advantage of the possibility of writing f(x) as

$$f(x) = \sum_{i=1}^{N} \alpha_i f_i(x) g_i(x) \tag{V.39}$$

where

$$\alpha_i \geq 0$$

$$0 \leq g_i(x) \leq 1$$

$$f_i(x) \geq 0 \tag{V.40}$$

$$\int_{-\infty}^{\infty} f_i(x)dx = 1.$$

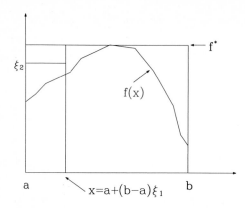

Fig.V.3 The "rejection technique" is applicable whenever the probability distribution function, f(x), has a finite extent. In the figure this is shown by enclosing f(x) in a rectangle of sides (b-a) and f^*. To select a value x according to the distribution f(x) one generates two random numbers, x_1 and x_2. The value $x=a+(b-a)x_1$ is accepted only when x_2 is smaller than f(x).

Then x is sampled as follows
1. Select i with probability proportional to α_i,
2. Select x from $f_i(x)$,
3. Accept x with probability $g_i(x)$,
4. If x is rejected return to 1.

The efficiency of this method is $1/\Sigma\alpha_i$ (Messel and Crawford, 1970) and since generally the decomposition, Eq(V.39), is not unique it becomes a matter of ingenuity to find the optimum set of functions $f_i(x)$, $g_i(x)$.

The techniques just described cover most of the cases encountered in practical applications. Quite often, however, it is possible to find ad hoc procedures which are faster (in terms of computing time); since Monte Carlo simulations involve very large numbers of such sampling operations the saving in running time may be substantial. As an example consider:

$$f(x) = 3\,x^2, \quad 0 \le x \le 1. \tag{V.41}$$

According to Eq(V.29) one obtains x from

$$x = \zeta^{1/3} \tag{V.42}$$

An alternative procedure is to select the largest of 3 random numbers, ξ_1, ξ_2, ξ_3, which is much faster than calculating the cubic root.

The extension of the methods described above to distributions of more than one random variable is straightforward.

V.2.2 Geometrical Randomness

The notion of a random variable X uniformly distributed in [a,b] is fairly intuitive. An extension of this concept to geometric objects uniformly and randomly distributed in Euclidean space is however not so obvious as different definitions of "randomness" are possible. A classical example is the famous Bertrand problem[18] (Kendall and Moran, 1963). Our interest here in this issue is related to the fact that, as charged particles are transported, one needs to generate random samples of segments along which the free flights (between consecutive interactions) occur. More generally, the view mentioned in Section V.1 that microdosimetric distributions result from the random overlap of two "objects", the track and the sensitive volume, requires again a consideration of this subject.

The following simple example illustrates the nature of the problem (Santalo, 1976). Assume we want to generate random points in a plane. We could use pairs of cartesian coordinates (x,y) or, for instance, polar coordinates (r,ϕ). Which one is right ? The answer depends on the definition of *measure* for a set of points in the plane. In the context of Euclidean geometry it appears desirable to ask that such a measure be invariant under the usual transformations, i.e. translations, rotations and reflections. The most general such transformation is

$$x' = a + x \cos \alpha - y \sin \alpha$$
$$y' = b + x \sin \alpha + y \cos \alpha$$

(V.43)

The measure $m(S)$ of the set of points S is

$$m(S) = \int_S f(x, y) dx dy$$

(V.44)

---

[18] The question raised by Bertrand was to find the probability that a random chord in a circle of unit radius has length greater than the side of an inscribed equilateral triangle (i.e. $\sqrt{3}$). The paradox rests with the fact that, apparently, it is possible to give three different answers to this question: a) select a point, A, on the circumference of the circle; if a second point, B, is selected randomly on the same circumference then the probability that $AB > \sqrt{3}$ is 1/3. b) on any given radius select a random point; then the probability is 1/2 that a perpendicular on the radius at this point is larger than $\sqrt{3}$. c) the centers of all chords larger than $\sqrt{3}$ must be within a circle of radius 1/2; since the total area of the circle is p the required probability (given by the ratio of the two areas) is 1/4. Under the assumption of μ-randomness (i.e. the circle is exposed to a uniform and isotropic field of infinite straight line) the correct answer is 1/2.

(or $\int f_1(r,\phi)drd\phi$, in polar coordinates) and the problem is to find $f(x,y)$ such that

$$m(S) = m(S') \tag{V.45}$$

or

$$\int_S f(x, y)dxdy = \int_{S'} f(x', y')d'xd'y \tag{V.46}$$

Note that the *same* function $f(x,y)$ is used on both sides of Eq(V.46). The Jacobian of Eq(V.43) is unity and the condition, Eq(V.46), amounts to

$$f(x, y) = f(x', y') \tag{V.47}$$

or, since x', y' are arbitrary

$$f(x, y) = \text{constant.} \tag{V.48}$$

Without any loss of generality we can set $f(x,y)=1$ to obtain:

$$m(S) = \int_S dxdy \tag{V.49}$$

To generate random points one selects (x,y) uniformly distributed or in polar coordinates $(dxdy=rdrd\phi)$ pairs of numbers (r^2,ϕ) uniformly distributed. Similarly, in three dimensional space random points are sampled with three uniformly-distributed numbers representing either (x,y,z) or $(r^3,\cos\theta,\phi)$. For instance, to generate a direction from an isotropic distributions one would use:

$$\cos\theta = 2\xi_1 - 1 \quad \phi = 2\pi\xi_2 \tag{V.50}$$

V.2.3 An Illustration: Monte Carlo Simulation of Electron Tracks

At the core of any charged-particle transport code is a routine for handling the interactions of electrons with the medium. Secondary electrons (also termed "delta rays") are the main vehicle by which energy is transported away from the trajectory of the primary track. There are several computer codes currently in use which transport charged particles *event-by-event*[19] using Monte Carlo techniques

---

[19] The term *event-by-event* refers to the fact that one records positions and energy transfers at each elastic or non-elastic collision event. A second method - generally inappropriate for microdosimetry but useful in dosimetric calculations - consists of dividing the medium into subvolumes, e.g. slabs, and compute only the total energy deposited by the track as well as new secondary particles resulting from the traversal of the slab. "Macroscopic" is taken to mean "large" relative to the mean free-path of the particle but "small" in relation to dimensions over which, due to changes in the particle energy, the cross sections will vary appreciably.

(Paretzke et al, 1974; Hamm et al, 1981; Zaider et al, 1983). These codes are essentially equivalent, although differences may exist concerning the treatment of particular processes; for instance, in some codes subexcitation electrons are not further followed while in others electrons are transported until they reach thermal energies. The decision to include such processes (which may require longer running times) depends on the kind of microdosimetric quantities that will be calculated: for instance, if the purpose of the calculation is to reproduce microdosimetric spectra obtained with a proportional counter (in which only ionizations are detected) one may safely ignore subexcitation electrons. On the other hand, the treatment of the interaction and diffusion of radical species induced by the charged particle in water is very sensitive to the final position of the electron: upon solvation (at thermal energies) this "aqueous" electron may diffuse over extensive distances before interacting (see Section VII.1).

The objective of a Monte Carlo simulation event-by-event is to obtain samples of particle trajectories (a trajectory is a collection of straight-line segments connecting consecutive elastic or non-elastic events). The following question is addressed in a typical Monte Carlo transport calculation: given the position and velocity of a particle at a certain time, what is the position of the next scattering center and the physical process responsible for the interaction ? The computing procedure which solves this problem (described below) is repeated sequentially for the primary as well as the secondary, higher-generation particles (delta rays) resulting from impact ionization. Particles - here electrons - are followed until they slow down to a preset threshold energy, or simply until a certain penetration depth has been obtained.

Let σ_i be the cross section for process "i" (e.g. excitation) and n the density of targets (scattering centers per unit volume). The probability of a type "i" scattering event along the infinitesimally short segment dx is

$$dp_i = n\,\sigma_i\,dx \qquad\qquad (V.51)$$

and from the definition of cross section, the probability of a free flight, x, followed by this type of scattering at dx about x is:

$$f(x)dx = e^{-\int_0^x n\sigma_T(t)dt}\, n\sigma_i\,dx \qquad\qquad (V.52)$$

where σ_T is the total interaction cross section:

$$\sigma_T = \sum_i \sigma_i. \qquad\qquad (V.53)$$

For an electron of energy E the path length to the next interaction is sampled with:

$$x = -\frac{1}{\sigma_T(E)n} \log(\xi)$$

(V.54)

The mean free path is (see Eq.III.18)

$$\bar{x}(E) = \frac{1}{\sigma_T(E)n}.$$

(V.55)

For water molecules at density $\rho=1$ g/cm^3, one has $n=334.3 \cdot 10^{20}$ molecules/cm^3. Thus at $E=1$ keV (see Fig.V.4) $\bar{x}=1.8$ nm. If only non-elastic interactions are considered then $\bar{x}=3.4$ nm. To put these numbers in some perspective one notes that at this density the average distance between two water molecules is 0.3 nm.

In the following we use the code DELTA (Zaider et al, 1983) to illustrate the implementation of Monte Carlo methods in transporting electrons in water vapor.

The *elastic* scattering of high-energy electrons (E>0.2 keV) may be described with the well-known Rutherford formula modified with a screening parameter, η (Bethe, 1953):

$$\sigma_e(E, \theta) = \frac{Z^2 e^4}{m^2 v^4 (1 + 2\eta - \cos\theta)^2}$$

(V.56)

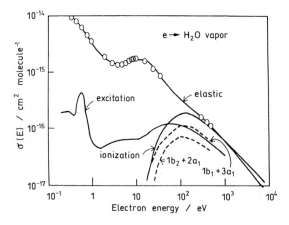

Fig.V.4 Calculated (lines) and measured (circles) cross sections for the interaction of electrons with water vapor as a function of electron energy (after Zaider et al, 1983).

or, integrated over the solid angle:

$$\sigma_e(E) = \frac{\pi\,Z^2\,e^4}{4E^2\eta(\eta+1)}. \qquad (V.57)$$

Here $\sigma_e(E,\theta)$ is the probability that an electron (kinetic energy E, velocity v, charge e and mass m) will elastically scatter in the solid angle $d\Omega=2\pi d(\cos\theta)$ in a medium of atomic number Z. The screening parameter is given by (Moliere, 1947):

$$\eta(E) = (\alpha_1 + \beta_1 \log E)\,\frac{kZ^{2/3}}{\dfrac{E}{mc^2}(E\,/\,mc^2 + 2)} \qquad (V.58)$$

where $\alpha_1=1.64$, $\beta_1=-0.0825$, $k=1.7\ 10^{-5}$.

Several empirical equations have been proposed for describing the elastic scattering of low-energy electrons. These expressions are generally chosen such that they will match asymptotically the Rutherford formulae, Eqs(V.56,57). One such set of equations was proposed by Porter and Jump (1978):

$$\sigma_e(E,\theta) \cong \frac{1}{\left[1 + 2\gamma(E) - \cos\theta\right]^2} + \frac{\beta(E)}{\left[1 + 2\delta(E) + \cos(\theta)\right]^2} \qquad (V.59)$$

$$\sigma_e(E) = T_1\left[\frac{E^x}{\dfrac{U}{E}\left(\dfrac{U}{E}+1\right)(V^{2+x} + E^{2+x})} + \sum_{n=1}^{2}\frac{F_n\,G^2_n}{(E - E_n)^2 + G^2_n}\right] \qquad (V.60)$$

Here $U=0.00195$, $x=-0.77$, $F_1=94$, $F_2=16.6$, $G_1^2=0.225$, $G_2^2=396.9$, $E_1=0.172$, $E_2=13.93$, $n=2$, $T_1=1.296$. These parameters have been obtained by fitting these expressions to available experimental data. Detailed expressions for the functions $\beta(E)$, $\gamma(E)$ and $\delta(E)$ can be found in Porter and Jump (1978).

Angle-integrated and double-differential cross sections for elastic scattering of electrons on water vapor are shown in Figs.V.4 and V.5, respectively, as calculated with Eqs(V.56-60). The experimental data shown for elastic scattering are from Seng and Linder (1976), Brüche (1929), Nishimura (1979) and Bromberg (1975). The experimental data shown for ionization (dashed curves) are from Märk and Egger (1976) and Schutten et al (1966).

To sample values $x=\cos\theta$ from the distribution function, Eq(V.59), one can use the techniques described above. However the method, Eqs(V.39-40), is more

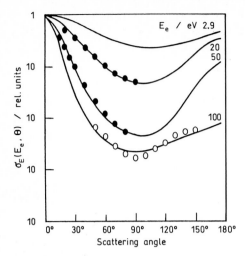

Fig.V.5 Differential cross sections for the elastic scattering of electrons (energy E_e) in water vapor as a function of scattering angle. Measured data are shown as circles (after Zaider et al, 1983).

convenient. One could factorize Eq(V.59) as follows:

$$\sigma_e \,(E,\,x) = \left[\frac{1}{2\gamma(1+\gamma)} \right]\left[\frac{2\gamma(1+\gamma)}{(1+2\gamma-x)^2} \right][1] +$$

$$+\left[\frac{\beta}{2\delta(1+\delta)} \right]\left[\frac{2\delta(1+\delta)}{(1+2\delta+x)^2} \right][1] \qquad (V.61)$$

The terms in square brackets represent $\alpha_1 f_1(x)g_1(x)$ and $\alpha_2 f_2(x)g_2(x)$, as indicated in Eq(V.39).

Impact *ionization* of the water molecule results in an electron removed from one of the five ground-state levels at -539.7, -32.4, -18.51, -14.75 or -12.62 eV. For low-energy electrons the experimental total ionization cross sections, $\sigma_i(E)$, of Märk and Egger (1976) and Schutten et al (1966) have been directly tabulated in the code; the selection of a particular orbital is made with the method of Eq(V.36). For higher energies (E>0.5 keV) the so-called Fano plot, $\sigma_i E$ versus $A + B\log(E)$, has been used to obtain a straight-line extrapolation of the experimental data at lower energies:

$$\sigma_i = \frac{0.167 \log(2E) + 0.505}{E}. \qquad (V.62)$$

In this equation the numerical constants correspond to E in keV and σ in units of 10^{-16} cm^2. Total ionization cross sections for electron impact on water vapor are shown in Fig.V.4.

The secondary-electron energy distribution, $\sigma_i(E,E')$, is represented in DELTA using the expression proposed by Green and Sawada (1972):

$$\sigma_i(E, E') = \frac{\Gamma(E)}{\left[E' - \Delta(E)\right]^2 + \Gamma^2(E)} \tag{V.63}$$

where E' is the energy of the secondary electron (by definition the faster electron is considered the primary). The empirical functions Γ and Δ are:

$$\Gamma(E) = \frac{12.8E}{E + 12.6},$$

$$\Delta(E) = 1.28 - \frac{1000}{E + 25.2}, \tag{V.64}$$

where E is in eV.

The angular distribution of the primary (θ,ϕ) and secondary (θ',ϕ') electrons is not known with any degree of accuracy. However, for higher energy secondary electrons one may use (Grosswendt and Waibel, 1978):

$$\sin^2 \theta = \frac{\dfrac{E'}{E}}{\left(1 - \dfrac{E'}{E}\right)\dfrac{E}{2\,mc^2} + 1}$$

$$\sin^2 \theta' = \frac{1 - \dfrac{E'}{E}}{1 + \dfrac{E'}{2\,mc^2}} \tag{V.65}$$

with ϕ randomly distributed in $[0,2\pi]$ and $\phi'= \phi-\pi$. These expressions have been found in reasonable agreement with the experimental evidence. For lower energies (E'<50 eV) it has been suggested (Grosswendt and Waibel, 1978) that secondary electrons are emitted isotropically while at intermediate energies (50 eV$\leq$E'$\leq$200 eV) some 90% of the secondary electrons are emitted in $[\pi/4,\pi/2]$.

Although the treatment of the angular distribution of the low-energy secondary electrons is rather imprecise, the impact this may have on the spatial structure of the track is expected to be minor since - as already indicated - at low electron energies elastic scattering (and not ionization or excitation) is the dominant process.

There are 12 known *excitation* levels in the water molecule. Table.V.1 lists their energies, states and description.

Table V.1

E/eV	State and description
0.198	$(A_1)v_2$
0.46	$(A_1)v_1+B_1v_3$
0.898	$(B_1)v_1+(B_1)v_3$
7.44	A^1B_1
9.85	B^1A_1
13.5	$1b_2 \rightarrow 3s$
16.95	H^* Lyman α
18.73	H^* Balmer α
9.5	OH^* (3064 Å)
9.998	$A + B$
11.057	$C + D$

At primary electron energies below about 10 eV vibrational excitations are dominant. As the energy increases other excitation channels open. Fig.V.6 is a

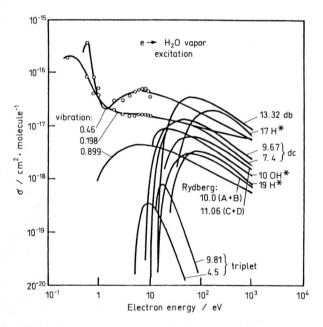

Fig.V.6 Excitation cross sections for the interaction of electrons with water vapor as a function of electron energy (after Zaider et al, 1983).

graphical compilation of data and semi-empirical interpolation. The cross sections were represented with an expression originally suggested by Green and Stolarski (1972):

$$\phi_{ex}(E) = \frac{A}{W^2}\left(\frac{W}{E}\right)^{\Omega}\left[1 - \left(\frac{W}{E}\right)^{\beta}\right]^{\nu},$$

(V.66)

where W is the excitation energy (see Table 1) and A, Ω, β and ν are W-dependent parameters (for more details see Zaider et al, 1983). For each excitation energy, W, a Fano plot was used to obtain cross sections at E>1000 eV. Following impact excitation no angular deflection of the incident electron is assumed.

A simulated 1-keV electron track is shown in Fig.V.7.

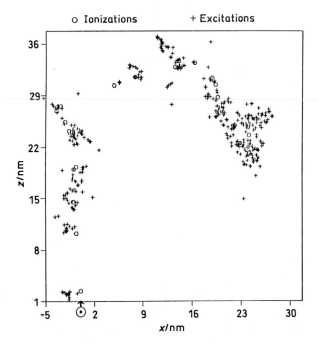

Fig.V.7 Projection in the (x-z) plane of a simulated 1-keV electron track. The track is started at the origin (x=y=z=0) in the positive z direction, as indicated in the figure by an arrow. Circles and crosses indicate ionizations and excitations, respectively. (after Zaider et al, 1983).

V.3 Calculation of Microdosimetric Spectra

V.3.1 Analytic Methods

By and large, analytic methods for calculating microdosimetric spectra are of only limited usefulness. It is generally necessary to make rather drastic approximations in the description of the charged particle tracks, most often the so-called *track segment approximation* where the charged particle trajectory in the medium is assumed to be a straight line segment. As a simple illustration, consider charged particles represented by infinitely long straight lines of constant LET (=L). For a spherical site of diameter d the resulting microdosimetric spectrum is given by the distribution of the quantity Lx, where x is a chord (path length) along which the particle traverses the site. One may thus identify the spectrum with the distribution of cord lengths in a sphere. The result, as shown in Chapter IV, is the well known "triangular" distribution:

$$f_1(z) = \frac{2z}{z_0^2}, \quad z \in [0, z_0] \tag{V.67}$$

The maximum specific energy, z_0, is given by:

$$z_0 = \frac{6}{\pi\rho\, d^2}\, L \tag{V.68}$$

This equation may be understood as follows: The specific energy deposited by a particle traversing a spherical site (volume v, density ρ) along the chord x is:

$$z = \frac{xL}{\rho v} = \frac{xL}{\rho \frac{4}{3}\pi(d\,/\,2)^3} \tag{V.69}$$

The distribution in z is identical (within a normalization constant) with the chord length distribution:

$$p(x) = \frac{2x}{d^2} \quad x \in [0, d] \tag{V.70}$$

Let C be the normalization factor. Then:

$$f_1(z) = Cz \quad z \in [0, z_0] \tag{V.71}$$

where z_0 is the maximum value of z (clearly, for a particle traversing the sphere along the diameter). C may be obtained from the normalization requirement:

$$I = \int_0^{z_0} f_1(z)\,dz \;=\; C\int_0^{z_0} z\,dz \;=\; \frac{C\,z_0^2}{2} \qquad\qquad \text{(V.72)}$$

Thus one regains the result, Eq(V.67). Fig.V.8 (a and b) gives an example of triangular microdosimetric distribution; two representations, $f(z)$ vs z and $z^2 f(z)$ vs $\log(z)$, are shown (compare with Fig.IV.6).

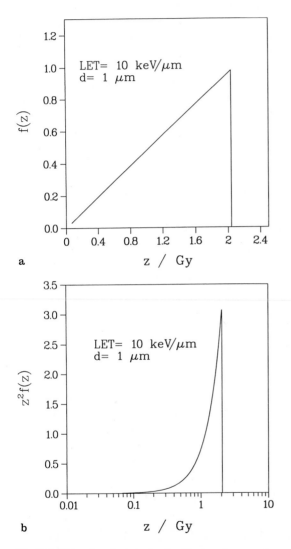

a

b

Fig.V.8 The microdosimetric distribution for a track segment (here of 10 keV/mm) in a spherical cavity has a triangular shape (Fig.V.8a). In the representation of Fig.V.8b, $z^2 f(z)$ vs $\log(z)$, the area under the curve is proportional to the fraction of the dose delivered in each z interval.

A more sophisticated version of this approach has been used for calculating analytically microsimetric spectra for neutrons. This method (Caswell and Coyne, 1974) is frequently used because of its relative simplicity and is described in detail in the following. This allows us to introduce certain other quantities of interest in radiation physics.

Under certain simplifying assumptions (to be specified below) it is possible to calculate the distribution of energy deposited in a site (volume V, surface area A) surrounded by an infinite medium of similar chemical composition and exposed to a uniform flux of neutrons with the aid of *transport theory*. This distribution is the microsimetric spectrum of the neutron field.

Let $S(r,E,\Omega,t)dr^3dEd\Omega$ be the expected number of charged particles produced per unit time in d^3r with energy E in the interval [E, E+dE], moving in the direction $d\Omega$ about Ω. The quantity S is the "source" spectrum of the neutron-induced charged particles. A quantity such as S will be called the source *distribution in E and* Ω. Let further $n(r,E,\Omega,t)d^3rdEd\Omega$ denote the distribution of the *expected* (i.e. average) number of charged particles in the volume d^3r about r with respect to E and Ω. We shall demonstrate that knowledge of these two quantities, S and n, allows the calculation of the microsimetric spectra.

The source and the particle density distributions are, obviously, not independent. The link between them can be given in terms of a *transport equation* which is basically a formal enumeration of all known paths by which particles can leave or enter the phase space volume element $dEd\Omega$. Thus

$$\frac{\partial n}{\partial t} + \mathbf{\Omega}\nabla(vn) + \mu_T(vn) =$$

$$\int_0^\infty dE' \int_{4\pi} d\Omega'(E' \to -E, \Omega' \to \Omega)v'n(r, E', \Omega', t) + S \qquad (V.73)$$

Here v is the velocity of a particle of energy E, $v\mu_T(r,E)$ is the number of interactions, per unit time, experienced at r by a particle with energy E; and $\mu_S(r,E \to E',\Omega \to \Omega')dE'd\Omega'$ is the probability per unit distance traveled that a particle with E and Ω will produce, as a result of an interaction at r, one or more other particles (including the incoming particle) with E' and Ω'. Following these definitions:

$$\mu_T(r, E, \mathbf{\Omega}) = \int_0^\infty dE' \int_{4\pi} d\Omega' \, \mu_s(r, E \to E', \mathbf{\Omega} \to \mathbf{\Omega'}). \qquad (V.74)$$

An elementary derivation of Eq(V.73) can be found in Morse and Feshbach (1953).

We make now a number of simplifying assumptions; they should be clearly understood, as the applicability of the expressions obtained below depends on their validity.

Firstly, assume that we are dealing with a *stationary* (that is time-independent) process. We can integrate out the time variable and use, instead of n, the new quantity

$$\psi(r, E, \Omega) = \int vn(r, E, \Omega, t)\, dt. \tag{V.75}$$

which is the charged-particle *fluence* distribution in E and Ω. Eq(V.73) becomes:

$$\Omega \Delta \psi + \mu_T \psi = \int dE' \int d\Omega'\, \mu_S (E' \rightarrow E, \Omega' \rightarrow \Omega)\psi(r, E', \Omega') + S(r, E, \Omega). \tag{V.76}$$

The *second* assumption we are making is known as *charged particle equilibrium* and postulates that the source, S, is uniformly distributed, i.e. it is independent of *r*. This further simplifies Eq(V.76) since (a) ψ does not depend on *r* and the gradient term is zero, and (b) one can integrate out Ω. Now

$$\mu_T (E)\psi(E) = \int_E^{\infty} d'E\, \mu_S (E' \rightarrow E)\psi(E') + S(E). \tag{V.77}$$

We are now making the *third* assumption, namely that energy loss by the charged particles is a continuous (rather than discrete) process. In terms of the stopping power of the medium, dE/dx, this so-called *continuous slowing down approximation* (CSDA) can be expressed as:

$$\mu_S (E' \rightarrow E) = \frac{1}{\Delta x}\, \delta\left[(E' - E) - \frac{dE}{dx}\, \Delta x\right] \tag{V.78}$$

Using Eq(V.74) for μ_T and integrating Eq(V.77) over E from E_0 to infinity, one obtains

$$\int_{E_0}^{\infty} dE\psi(E) \int_0^E dE'\, \mu_s (E \rightarrow E') =$$

$$= \int_{E_0}^{\infty} dE \int_E^{\infty} dE'\, \mu_s (E' \rightarrow E)\psi(E') + \int_{E_0}^{\infty} S(E)dE \tag{V.79}$$

The double integral on the right-hand side of this expression can further be written as

$$\int_{E_0}^{\infty} dE' \int_{E_0}^{E'} dE\, \mu_s (E' \rightarrow E)\psi(E'), \tag{V.80}$$

which, upon reversing notation (E→E', E'→E) makes Eq(V.79) take the following form:

$$\int_0^{E_0} dE' \int_{E_0}^{\infty} dE\, \psi(E)\, \mu_s(E' \to E) = \int_{E_0}^{\infty} S(E)dE \tag{V.81}$$

Finally, using CSDA [Eq(V.78)], one obtains

$$\int_0^{E_0} dE \int_{E_0}^{\infty} dE\, \psi(E)\, \frac{1}{\Delta x}\, \delta\left[E - \left(E' + \frac{dE}{dx}\, \Delta x \right) \right]$$

$$= \frac{1}{\Delta x} \int_{E_0 - \frac{dE}{dx}\Delta x}^{E_0} dE'\, \psi(E' + \Delta x) \tag{V.82}$$

or, in the limit $\Delta x \to 0$:

$$\psi(E_0)\, \frac{dE}{dx} = \int_{E_0}^{\infty} S(E)dE. \tag{V.83}$$

The expression, Eq(V.83), is the desired result which allows to calculate the so-called slowing-down particle distribution, $\Psi(E)$, when the source distribution in energy is known. The derivation of Eq(V.83) given above follows largely that of Roesch (1968).

In calculating the microdosimetric spectra it is useful to make, as Caswell and Coyne (1974) have done, a *fourth* assumption, namely that charged particles travel in straight lines and have no radial extension in terms of energy deposition. This assumption leads to a direct link between the geometry of the volume exposed to the radiation field and the energy deposited.

Let f(e)de be the number of charged particles depositing in a volume energy in the interval [e,e+de]. For particles coming from *outside* the volume:

$$f_1(e)de = \int_{R(e)}^{d} [\psi(E) + \psi(e)]\, f_\mu(x)dx. \tag{V.84}$$

In this equation R(E) is the **range** of a charged particle of energy E (see section III.6.3):

$$R(E) = \int_0^E \frac{dE'}{dE' / dx} \tag{V.85}$$

and $f_\mu(x)dx$ is the fraction of chords of length x to x+dx when the volume is exposed to a uniform and isotropic fluence of straight lines. The expression,

Eq(V.84), can be understood as follows: along each chord of length x one can deposit the energy, e, in two different ways:

(a) from a charged particle which has energy E upon entering the volume, and energy E-e at the exit, i.e.

$$x = R(E) - R(E - x);$$
(V.86)

(b) from a charged particle with energy e and range such that $R(e) \leq x$.

The square brackets in Eq(V.84) contain the total fluence of such particles. Particles in categories (a) and (b) are called *crossers* and *stoppers*, respectively.

For particles "born" inside the volume one has, using similar arguments:

$$f_2(e)de = \overline{x}\int_{R(e)}^{d} [S(E) + S(\varepsilon)] f_i(x)dx$$
(V.87)

Here $f_i(x)$ refers to the distribution in length, x, of isotropic rays originating in random points in the volume. The two integrals in Eq(V.87) count *starters* and *insiders*, respectively. The energy E satisfies Eq(V.85), as in the previous case, and x is the mean chord length in the volume. Finally,

$$f(e) = f_1(e) + f_2(e)$$
(V.88)

The two chord-length distributions are not independent. It can be shown (Kellerer, 1971, 1985) that

$$f_i(x) = \int_x^\infty f_\mu(x')d'x$$
(V.89)

Formulae for $f_\mu(x)$ for spheroids and rectangular parallelepipeds have been obtained by Kellerer (1985) and Coleman (1973).

V.3.2 Monte Carlo Methods

The net result of a Monte Carlo transport calculation is a simulated track consisting of the geometrical positions of all energy deposition events as well as the amount of energy deposited at each interaction point. Fig.V.7 shows an example of a simulated track. The calculation of microdosimetric spectra may be performed with the aid of Monte Carlo techniques. In an experimental determination of $f_1(z)$ a fixed volume (the proportional counter) is traversed by random tracks and the energy deposited recorded event by event. The calculation adopts the opposite procedure, namely, on one or several tracks large numbers of sampling volumes are randomly placed, and each time the total energy deposited is stored. Naively, one could think of defining a sufficiently large box completely enclosing the track and then throwing sampling volumes (say, spheres) randomly into the box. The efficiency of this procedure, measured as the ratio of successes (spheres containing at least one energy-transfer event) to the total number of spheres sampled, is generally less than one. "Missing" may be in fact rather costly in terms of computing time since verifying the content of each sphere involves a loop over a large number of transfer points. It is therefore desirable to bring the efficiency as close to one as possible. This is particularly important for very small sites. A more efficient method for obtaining microdosimetric spectra is to use the concept of *associated volume* introduced earlier.

The associated volume of a track is the volume around the track that has a sampling efficiency of exactly one. Conceptually it can be built as follows (see Fig.V.9): Centered at each transfer point place a sampling sphere. The union of all these spheres is the associated volume. It has the property announced since, by its construction, any sphere placed with its center in it will be within at most one radius from at least one transfer point. For simple geometric volumes the associated volume may be calculated directly. For instance, for a track segment of length l (with no radial extension and treated within the continuous slowing down approximation) we have:

$$V = r^2 \pi l + \frac{4\pi r^3}{3} \qquad (V.90)$$

where V is the associated volume and r is the radius of the sampling sphere.

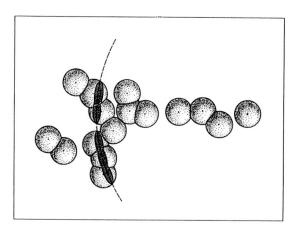

Fig.V.9 The problem described in this figure is that of finding a volume such that whenever a given site (here a sphere) is placed with the center in this volume one is insured to have at least one energy transfer point inside the site. To obtain this associated volume of the track one takes the union of all sites (spheres) centered at energy transfer points. The property announced follows immediately. The associated surface is the intersection of the associated volume with a sphere of radius R centered at the origin of the track, here the left-most energy transfer.

If we denote with v the volume of the box then the sampling efficiency is V/v. This ratio depends on the radius r and clearly:

$$\lim_{r \to 0} \frac{V}{v} = 0 \qquad\qquad (V.91)$$

For low LET radiation and r<10 nm the efficiency is typically lower than 1%.

A computational scheme for directly simulating the associated volume, that is for selecting spheres in the AV only, may proceed as follows: a) select randomly a transfer point, b) within a distance r of this point place randomly a sphere of radius r, and c) calculate the total energy in this sphere. The event thus obtained should be scored in the spectrum with a weight inversely proportional to the number of events in the sphere; the reason for this is that this method is biased towards regions of high density of transfer points (sampling should be spatially uniform).

A key element in this method (and also in the definition of the associated volume) is that tracks intersecting the sphere originate uniformly from random points in space and are isotropically distributed. This has been called μ *randomness* (Kellerer, 1985). In the calculation μ randomness means that spheres are selected homogeneously in space. There are situations, however, where this conditions are not satisfied. An important example (Zaider and Varma, 1990) is exposure to radon alpha particles that occurs under what might be called *modified surface randomness* (designated as σ-*randomness*). In this case bronchial cells in the lung are exposed to isotropic distributions of a particles originating from a point at fixed distance, R, from the center of the cell (site). The unit-efficiency locus is now the intersection between the associated volume and the sphere of radius R centered at the origin of the track (see Fig.V.9). By analogy with the AV one may term this *associated surface* (AS) of the track. The simulation of the AS is similar to that of the AV (the geometry of the problem is however more complicated). The main difference is in calculating the bias introduced by this method: the selection of spheres is a) proportional with the number of transfer points in the sphere that are within [R-r,R+r] from origin, and b) inversely proportional to the AS zone contained in the sphere. An example of microdosimetric spectrum calculated with this approach is given in Fig.V.10.

V.3.3 Microdosimetric Spectra for Combined Radiations

A radiation field is generally a mixture of different radiations. For instance, in a neutron field one finds both neutron and gamma radiation. The questions arises as to how to obtain the microdosimetric spectrum for such a field when the microdosimetric distributions for the constituents are known.

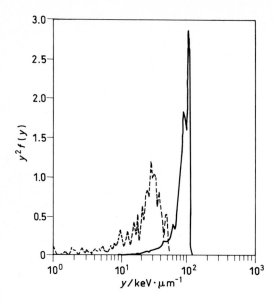

Fig.V.10 Microdosimetric spectra for a 7-MeV alpha particle that originates at a fixed distance (3 μm) from the center of the sphere. The two spectra shown are for spheres of 1 mm (solid line) and 10 nm (dashed line) diameter.

Let $f_i(y)$ be the microdosimetric spectrum for radiation i (i=1,2,...). Assume that D_i is the dose contributed by radiation i to the total dose delivered, D, and let further $z_F(i)$ represent the frequency-averaged specific energy of this radiation in a given volume. The microdosimetric distribution for the whole field, $f(y)$, is given by:

$$f(y) = \frac{\sum_i \dfrac{D_i}{z_F(i)} f_i(y)}{\sum_i \dfrac{D_i}{z_F(i)}} \qquad (V.92)$$

This is understood since the contribution of the spectrum $f_i(y)$ is proportional with the number of microdosimetric events delivered by that radiation.

As an application of this formula consider microdosimetric spectra for indirectly ionizing radiation (neutrons, photons). We shall use photons as an example. Let $n(E_\gamma)dE_\gamma$ be the number of photons in the field with energy in the interval $[E_\gamma, E_\gamma + dE_\gamma]$ at a total absorbed dose D. The absorbed dose contributed by these photons is:

$$D(E_\gamma) \propto n(E_\gamma) E_\gamma \mu(E_\gamma) / \rho \qquad (V.93)$$

Here μ is the linear attenuation coefficient and ρ is the density of the material. If for each energy E_γ we know the microdosimetric distribution, $f(y,E_\gamma)$, then for the full photon field one has:

$$f(y) \propto \int_0^\infty \frac{n(E_\gamma)\, E_\gamma\, \mu(E_\gamma)}{y_F(E_\gamma)} f(y, E_\gamma)\, dE_\gamma \qquad (V.94)$$

For convenience we have replaced z_F with y_F.

Assume now that we have distributions $f(y,E_e)$ for a given electron energy, E_e. Let $n(E_\gamma,E_e)$ be the primary electron spectrum for photons of energy E_γ. Then, similarly to the equation above:

$$f(y, E_\gamma) \propto \int_0^\infty \frac{n(E_\gamma, E_e)\, E_e}{y_F(E_e)} f(y, E_e)\, dE_e \qquad (V.95)$$

In these derivations we have assumed charged particle equilibrium.

We illustrate this procedure for the case of 200 kVp x rays. The x ray energy spectrum has been taken from Johns and Cunningham (1983) to correspond to a

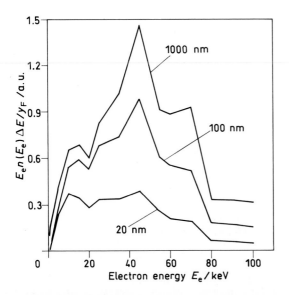

Fig.V.11 To obtain the microdosimetric spectrum of a photon field one takes a linear combination of microdosimetric distributions corresponding to the individual electrons set in motion by the photons. The coefficients of this linear combination are plotted here as a function of electron energy and site diameter. This particular example is for 250 kVp x rays. The numerical values on the ordinate are in arbitrary units.

half value layer of 3 mm of copper. The corresponding electron spectrum has been calculated with the aid of the computer code PHOEL2 (Turner et al, 1980) which takes into account photoelectric and Compton scattering of the photons. It should be noticed from Eq(V.95) that to obtain the microdosimetric spectrum of this field one uses a weighted average of spectra, $f(y,E_e)$, with the weight given by $E_e n(E_\gamma,E_e)/y_F(E_e)$; this quantity is plotted as a function of the electron energy in Fig.V.11 for 3 different diameters (20, 100 and 1000 nm) and it shows the relative importance for the final distribution of different electron energies.

The three curves are different since y_F depends on the site size. Since spectra for monoenergetic electron are available necessarily at a finite number of discrete energies it is important to select judiciously these energies. Fig.V.12 shows a comparison between a calculated spectrum for 250 kVp x rays and a measured distribution at d=1000 nm.

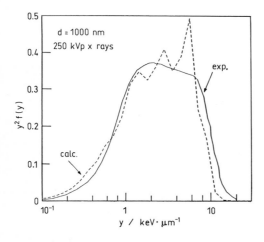

Fig.V.12 Calculated (dashed line) and measured (solid line) microdosimetric spectra in a 1μm diameter tissue equivalent sphere exposed to 250 kVp x rays. The measured spectrum was obtained by Braby and Ellett (1971).

V.4 Methods for Obtaining Proximity Functions

V.4.1 Proximity Functions for Simple Geometric Objects

In this section we provide mathematical expressions for proximity functions describing elementary geometric objects. Other than helping familiarize the reader, by way of simple examples, with the concept of proximity function, some of the expressions in this section also provide reasonable first approximations to proximity functions for tracks that resemble these geometric shapes. In the following we shall think of a track as a solid object of a certain shape and of uniform density. For instance, within the continuous slowing down approximation we may take a straight line segment to represent a charged particle track interacting with a medium. Or, when the radial extension of the track is to be included, the same may be achieved with a cylindrical shape. It will not be necessary to require that the volume (or surface or length, depending on the dimensionality of the problem) be finite; and also, the object may consist of a number of disjoint parts.

We recall the definition of proximity function[20]: given a random reference point in the object we consider the volume (number of points) of this object contained in a shell of radius x and thickness dx centered at the reference point. The average (expected) value of this quantity is the proximity function, t(x)dx. The expected volume contained in a sphere of radius x (centered at a random reference point) is denoted by T(x). These two quantities stand in the same relation as the differential and integral probability distributions, namely:

$$T(x) = \int_0^x t(x')d'x \qquad (V.96)$$

For objects of finite volume, V, an equivalent definition is:

$$t(x) = V p(x) \qquad (V.97)$$

where p(x)dx is the (normalized) probability distribution of distances between points randomly selected inside the object; p(x)dx is referred to as "point pair distance distribution".

Perhaps the simplest example of proximity function is that corresponding to an object occupying the whole space. Clearly:

$$t(x)dx = (4\pi x^2)dx \qquad (V.98)$$

We indicate elsewhere (Section V.4.3.2) that for any object t(x) can not increase faster than x^2. One may think of Eq(V.98) as representing a first approximation to

---

[20] The proximity function of energy transfer, defined in section V.1, refers to energy rather than volume; it may be obtained from the geometric proximity function defined below by multiplying this latter quantity with a factor representing the average energy per unit volume.

the proximity function of energy deposition for a low-LET (e.g. x ray) field where energy transfer points are quasi-randomly distributed in space.

Consider next a track consisting of a straight line of infinite length. In this case:

$$t(x)dx = 2dx \qquad (V.99)$$

If the track has finite length, L, one obtains by direct integration the result:

$$t(x)dx = 2\left(1 - \frac{x}{L}\right)dx, \quad x \in [0, L] \qquad (V.100)$$

The expression Eq(V.99) approximates the proximity function of a charged particle track (after multiplication by the LET of the track) if: a) x is much larger than the radial extension of the track, b) x is significantly smaller than the range of the particle, and c) over distances of the order of x the LET of the charged particle does not change. Where condition b) does not apply one can not use Eq(V.100) instead since, by definition, condition c) will not be satisfied either.

For a sphere of diameter d the proximity function is (NCRP, 1991):

$$t(x) = 4\pi x^2\left(1 - \frac{3x}{2d} + \frac{x^3}{2\,d^3}\right), \quad x \in [0, d] \qquad (V.101)$$

The generalization of this expression for a spheroid is (Kellerer, 1984):

$$
\begin{aligned}
t(x) &= U_1(x) && if\ e < 1,\ x \le ed \\
&= U_2(x) && if\ e < 1,\ x > ed \\
&= U_1(x) && if\ e > 1,\ x < d \\
&= U_1(x) + U_2(x) && if\ e > 1,\ x > d
\end{aligned}
\qquad (V.102)
$$

where two axes of the spheroid are d and the third axis is e*d, and:

$$U_1(x) = 4\pi x^2\left[1 - \frac{3x}{2d}\frac{c_1}{e} + \frac{x^3}{2\,d^3}\frac{c_2}{e}\right]$$

$$U_2(x) = 4\pi x^2\,\frac{3}{8}\,\frac{\varepsilon}{\frac{1}{e} - e}\left[\sqrt{\frac{d^2}{x^2} - 1}\left(\frac{x^2}{2\,d^2} + 1\right) + \left(\frac{x^3}{2\,d^3} - \frac{2x}{d}\right)ci\left(\frac{d}{x}\right)\right].$$

$$(V.103)$$

The following notations are used in these expressions:

$$\varepsilon = \sqrt{|e^2 - 1|}$$

$$c_1 = \frac{1}{2} + \frac{e^2}{2\varepsilon} ci\left(\frac{1}{e}\right)$$

$$c_2 = \frac{1}{4\, e^2} + \frac{3}{4} c_1 \qquad\qquad (V.104)$$

$$ci(x) = \cos^{-1}(x), \quad x \in [0,1]$$

$$= \cosh^{-1}(x), \quad x > 1$$

For a right cylinder with circular cross section (diameter d) of height h (Kellerer, 1981):

$$t(x) = (4\pi x^2)\,\frac{1}{x}\int_{z_1}^{z_2}\left(1 - \frac{z}{h}\right) U_c\left(\sqrt{x^2 - z^2}\right) dz,$$

$$\qquad\qquad (V.105)$$

$$x \le \sqrt{h^2 + d^2}, \quad z_1 = \sqrt{\mathrm{Max}(0,\, x^2 - d^2)}, \quad z_2 = \mathrm{Min}(x, h)$$

with:

$$U_c(x) = \frac{2}{\pi}\left[\cos^{-1}\left(\frac{x}{d}\right) - \frac{x}{d}\sqrt{1 - \frac{x^2}{d^2}}\right], \quad x \le d \qquad\qquad (V.106)$$

V.4.2 Proximity Functions for Amorphous Tracks

For charged particle tracks of constant LET the calculation of proximity functions may be greatly simplified by using the concept of amorphous track, introduced by Kellerer and Chmelevsky (1975). The requirement of "constant LET" is equivalent to evaluating the proximity function $t(x)$ only at small values of x, that is over distances where the LET of the track does not change significantly.

The interaction of a charged particle track with matter results in primary excitations and ionizations, these latter generating secondary electrons (δ rays). To simplify the terminology, let the primary energy transfer events (excitations and ionizations) be also considered as δ rays, either singly (for excitation) or as part of the secondary electron (following ionization). The track consists now of energy transfer points grouped in statistically independent δ rays. According to its definition, to evaluate the proximity function one selects randomly a reference energy transfer point in the track[21] and then evaluates the total energy at distance x to x+dx by adding contributions from all transfer points thus selected. Transfer points in the shell (x,x+dx) may belong to the same δ ray as the reference point or to other, uncorrelated δ rays. We denote the proximity functions for these two kinds of transfer points by $t_\delta(x)$ and $t_a(x)$, respectively. Clearly,

$$t(x) = t_\delta(x) + t_a(x) \qquad (V.107)$$

The proximity function for δ rays depends only on the energy spectrum of secondary electrons, f(E). Within a good approximation f(E) depends only on the velocity of the charged particle. Denoting the proximity function of an electron of initial energy E by $t_\delta(x,E)$ one has:

$$t_\delta(x) = \frac{\int\limits_0^\infty t_\delta(x, E)Ef(E)dE}{\int\limits_0^\infty Ef(E)dE}. \qquad (V.108)$$

The reason proximity functions for electrons of different energies, E, are combined with weight proportional to Ef(E) [rather than f(E)] is the following: assume a radiation field consisting of a large number of electrons with energies distributed according to f(E). To obtain the proximity function of the entire field, t(x), one selects reference points with probability proportional to the energy absorbed there. Transfer points belonging to an electron of initial energy E (E is, obviously, the sum of all energy transfers for this particle) will be therefore selected with probability proportional to E.

The second proximity function, $t_a(x)$, depends only on the average radial distribution of energy around the track (hence the term "amorphous" track). This is because energy transfer points included in this proximity function are not correlated with the reference point, and their contribution may be averaged out irrespective of this latter. Let g(b)db be the expected (average) fraction of energy deposited in a cylindrical shell of radius b and thickness db whose axis is the track core. A related quantity is G(b) defined as :

$$G(b) = \int\limits_0^b g(b')d'b \qquad (V.109)$$

---

[21] As a reminder, "random" means here proportional to the energy transferred locally.

g(b) may be used to obtain the dose, D(b), at distance (impact parameter) b from the track:

$$D(b) = \frac{L\,g(b)}{2\pi b \rho} \qquad\qquad (V.110)$$

where L is the linear energy transfer of the track and ρ is the density of the medium. The expression that relates $t_a(x)$ to g(b) may be obtained as follows: In Fig.V.13 the axis of the track is along z and O_1 represents a random point at impact parameter b with respect to the track. We take O_1 as reference point and consider the energy deposited in a sphere of radius x centered at O_1. To this end we need the fraction of G(b) contained in the sphere. The element of cylindrical surface, dS, on the intersection between a cylinder of radius b' and the sphere is:

$$dS = b'\,d\phi\,dz \qquad\qquad (V.111)$$

The total cylinder surface contained in the sphere is obtained by integration. The maximum value of the angle ϕ is (see Fig.V.14):

$$\phi_{max} = \arccos\frac{b'^2 + b^2 - x^2}{2b'b} \qquad\qquad (V.112)$$

For a given ϕ the integration range of z is from 0 to z_{max}. Thus:

$$S = 4b' \int_0^{\phi_{max}} d\phi \int_0^{z_{max}} dz \qquad\qquad (V.113)$$

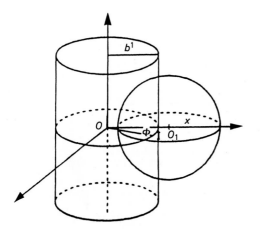

Fig.V.13 A sphere of radius x centered at O_1 intersects an amorphous track oriented along the z axis. The notations in the figure are used to derive Eq(V.115).

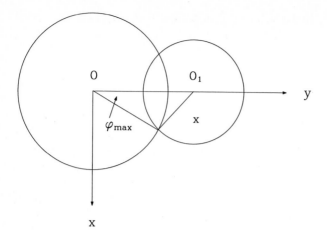

Fig.V.14 Projection on the (x,y) plane of the bodies shown in Fig.V.13.

where:

$$z_{max} = \sqrt{x^2 - b'^2 - b^2 + 2b'b \cos \phi} \tag{V.114}$$

To obtain the proximity function, $T_a(x)$, one needs to integrate over $g(b')$ (at a fixed b) and then over $g(b)$ for all possible positions of the reference point O_1. The integration limits are as follows (see Fig.V.13):

1) when $b>x$ then :
   $b'\in[b-x,b+x]$
   $\varphi\in[0,\varphi_{max}]$

2) when $b\leq x$ then $b'\in[0,b+x]$:
   when $b'\in[0,x-b]$ then $\varphi\in[0,\pi]$
   when $b'\in[x-b,x+b]$ then $\varphi\in[0,\varphi_{max}]$

The expression for $T_a(x)$ is:

$$T_a(x) = \frac{2}{\pi} \{ \int_0^x dbg(b) \int_0^{x-b} db'g(b') \int_0^{\pi} \sqrt{x^2 - b^2 - b'^2 + 2b'b \cos \phi}\, d\phi +$$

$$\int_0^x dbg(b) \int_{x-b}^{x+b} db'g(b') \int_0^{\phi_{max}} \sqrt{x^2 - b^2 - b'^2 + 2b'b \cos \phi}\, d\phi + \tag{115}$$

$$\int_x^{\infty} dbg(b) \int_{b-x}^{b+x} db'g(b') \int_0^{\phi_{max}} \sqrt{x^2 - b^2 - b'^2 + 2b'b \cos \phi}\, d\phi \}.$$

By differentiation with respect to x one obtains $t_a(x)$.

For charged particles of equal velocity $t_a(x)$ is proportional to the LET, L (the shape of the energy spectrum of δ rays is the same but their average number per unit length is proportional to L). Symbolically one can write:

$$t(x) = t_\delta(x) + t_{1a}(x)L \qquad (V.116)$$

with an obvious notation for $t_{1a}(x)$. For particles of equal velocity $t(x)$ can be found by calculating only two functions and then applying the expression, Eq(V.116). This is exemplified in Fig.V.15: all particles considered have an

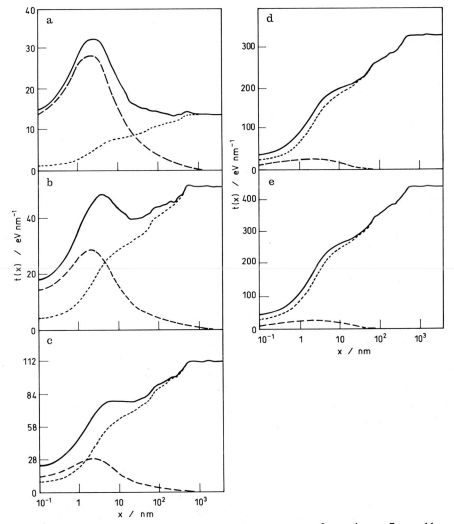

Fig.V.15 Proximity functions, t(x), for 6.58 MeV/amu ions: **a** ^{2}H, **b** ^{4}He, **c** ^{7}Li, **d** ^{11}B, **e** ^{12}C. Contributions from t_δ (long dash) and t_a (short dash) are shown separately (after Zaider and Rossi, 1985).

energy of 6.58 MeV/nucleon (and therefore the same velocity). We show separately the total proximity function, $t(x)$, and the contributions from $t_\delta(x)$ and $t_a(x)$. The particles are, from top to bottom, ^{2}H (6.5 keV/mm), ^{4}He (25 keV/mm), ^{7}Li (55 keV/mm), ^{11}B (165 keV/mm) and ^{12}C (220 keV/mm). It is interesting to notice that as the LET increases the proximity function is increasingly dominated by its amorphous component and the contribution of individual δ rays becomes negligible. This can be used as a criterion for approximating the particle with an amorphous track.

Because of the relation between y_D and $t(x)$ one can write similarly:

$$y_D = y_\delta + y_{1a} L \qquad (V.117)$$

It has been noted (Kellerer and Chmelevsky, 1975) that, since $t_\delta(x)$ does not depend on the angle at which the secondary electron is emitted, and since $g(b)$ can be obtained experimentally, the determination of $t(x)$ may not require knowledge on the angular distribution of the δ rays, an obvious advantage.

V.4.3 Proximity Functions from Experimental Data

Proximity functions of energy transfer, $t(x)$, are a convenient modality for representing in a compact form the geometrical pattern of energy transfer points in a charged-particle track. For simple geometrical structures the proximity functions may be evaluated analytically. In the simplest case, a linear track with no radial extension (i.e. all transfer points along the track) and constant linear energy transfer L one obtains, as has been indicated already:

$$t(x)dx = 2L\,dx \qquad (V.118)$$

Other expressions obtainable analytically have been given in Section V.4.1.

In Section V.1 we have introduced an important relation between the microdosimetric quantity, z_D, the dose-averaged specific energy and the two proximity functions, $t(x)$ and $s(x)$, characterizing respectively the charged particle track and the site where the microdosimetric spectrum is obtained:

$$z_D = \int \frac{t(x)\,s(x)}{4\pi x^2 m}\,dx, \qquad (V.119)$$

where m is the mass of the site. The importance of this relation rests with the fact that once $t(x)$ is known one may obtain z_D (a quantity useful in biophysical modeling, see Section VI.1.4) for *any geometric site* for which the proximity function, $s(x)dx$, is known. Conversely, if values z_D are known (e.g. from experiment) for a series of sites, $s_i(x)$, i=1,2..., one may attempt to solve the integral equation, Eq(V.119), for t(x) and thus relate the proximity function directly to a *measured* quantity. [By and large, proximity functions are obtained from Monte-Carlo-generated tracks].

An *analytic* solution for t(x) has been indeed obtained when s(x) refers to spherical sites of different diameters. This is particularly relevant since many of the proportional counters used to measure z_D are in fact spherical. A description of this particular solution is given below.

One may add that for those sites where no analytical solution obtains for t(x) numerical techniques may be used to invert Eq(V.119). This will apply, for instance, to cylindrical counters. The other pertinent factor in the application of these methods is the recent development of the variance-covariance technique (see section IV.5.2) for measuring z_D *without* the need to obtain first the full microdosimetric spectrum.

A second possibility for obtaining proximity functions from experimental data is to use tracks photographed in cloud chambers. In a cloud chamber each condensed droplet is associated with an ionization event; however one needs to account for the fact that droplets are affected by diffusion in the chamber. In Section V.4.3.3 we describe two procedures for "removing" the diffusional smearing of t(x) thus obtained.

V.4.3.1 t(x) and y_D

In this section we shall describe how to solve Eq(V.119) when s(x) represents a spherical site of diameter Δ:

$$s(x)dx = 4\pi x^2 \left(1 - \frac{3x}{2\Delta} + \frac{x^3}{2\,\Delta^3}\right)dx \qquad (V.120)$$

It is convenient to use the dose-averaged energy, e_D, instead of z_D. These two quantities are related:

$$z_D = e_D / m. \qquad (V.121)$$

With this, Eq(V.119) becomes:

$$e_D(\Delta) = \int_0^\infty u\left(\frac{x}{\Delta}\right)t(x)dx, \qquad (V.122)$$

where it is explicitly shown that a) the function $u=s(x)/4\pi x^2$ depends only on the ratio x/Δ, and b) e_D is a function of the site diameter, Δ. The problem is: given

$e_D(\Delta)$ and the kernel $u(x/\Delta)$, find $t(x)$. The solution may be obtained using the Mellin transforms.

Let $E(\sigma)$, $U(\sigma)$ and $T(\sigma)$ be the Mellin transforms of $e_D(\Delta)$, $u(x/\Delta)$ and $t(x)$, respectively. For instance:

$$T(\sigma) = \int_0^\infty x^{\sigma-1} t(x) dx,$$

(V.123)

$$t(x) = \frac{1}{2\pi i} \int_{C-i\infty}^{C+i\infty} x^{-\sigma} T(\sigma) d\sigma$$

(V.124)

where C is a positive real number selected such that there are no poles in the integrand. From Eq(V.122):

$$E(\sigma) = \int_0^\infty \Delta^{\sigma-1} d\Delta \int_0^\infty u\left(\frac{x}{\Delta}\right) t(x) dx$$

$$= \int_0^\infty x^{\sigma} t(x) dx \int_0^\infty y^{-\sigma-1} u(y) dy,$$

(V.125)

where $y = x/\Delta$. Thus [see Eq(V.123)]:

$$E(\sigma) = T(\sigma + 1) U(-\sigma)$$

(V.126)

or

$$T(\sigma) = E(\sigma - 1) / U(-\sigma + 1).$$

(V.127)

The expression, Eq(V.127), indicates that given the Mellin transforms, $E(\sigma)$ and $U(\sigma)$, one may calculate $T(\sigma)$ and then - with the aid of Eq(V.124) - its inverse, $t(x)$. $U(-\sigma+1)$ may be readily obtained by transforming $u(x/\Delta)$. The solution to Eq(V.122) is:

$$t(x) = \frac{1}{6\pi i} \int_{C-i\infty}^{C+i\infty} x^{-\sigma} (1 - \sigma)(2 - \sigma)(4 - \sigma) E(\sigma - 1) d\sigma.$$

(V.128)

This expression may be further simplified by using the (easily verifiable) statements:

If $F(\sigma)$ is the Mellin transform of $f(x)$ then:

a1) $F(\sigma+n)$ is the Mellin transform of $x^n f(x)$,
a2) $(-1)^n \sigma^n F(\sigma)$ is the transform of $(x\, d/dx)^n f(x)$,

where n is an integer. From a1) with n=1 and Eq(V.128) one obtains:

$$xt(x) = \frac{1}{6\pi i} \int_{C-i\infty}^{C+i\infty} x^{-\sigma} \sigma(\sigma - 1)(\sigma - 3)E(\sigma)d\sigma$$

$$= -\frac{1}{6\pi i} \int_{C-i\infty}^{C+i\infty} x^{-\sigma}(\sigma^3 - 4\sigma^2 + 3\sigma)E(\sigma)d\sigma \qquad \text{(V.129)}$$

Using a2) this becomes:

$$xt(x) = \frac{1}{3}\left[\left(x\frac{d}{dx}\right)^3 + 4\left(x\frac{d}{dx}\right)^2 + 3\left(x\frac{d}{dx}\right)\right]e_D(x) \qquad \text{(V.130)}$$

or

$$t(x) = \frac{1}{3}\left[x^2\frac{d^3}{dx^3} + 7x\frac{d^2}{dx^2} + 8\frac{d}{dx}\right]e_D(x). \qquad \text{(V.131)}$$

The notation used above for the derivatives, $(xd/dx)^n$ means (as an example, for n=2):

$$\left(x\frac{d}{dx}\right)^2 f(x) = x\frac{d}{dx}\left(x\frac{df}{dx}\right) = x\left(x\frac{df}{dx}\right)^2 + x^2\frac{d^2f}{dx^2}. \qquad \text{(V.132)}$$

The practical application of Eq(V.131) is not as straightforward as its derivation. The evaluation of the first three derivatives of $e_D(x)$ from a discrete set of experimental points, $e_D(x_i)$, i=1,2,..., requires very accurate data. It is always easier to obtain, if possible, an analytic approximation to the data over a range of x values. We quote in this respect data by Bengtsson and Lindborg (1974) who obtained experimental values of the dose-averaged lineal energy, y_D, for a ^{60}Co beam within the range of diameters of 11 nm to 22 μm. As a reminder

$$e_D = \frac{2\Delta}{3} y_D. \qquad \text{(V.133)}$$

For these data Forsberg et al (1978) have suggested the function:

$$y_D(\Delta) = 2.01\ \Delta^{-0.4,} \qquad \text{(V.134)}$$

where y_D and Δ are respectively in keV/μm and μm. When introduced in Eq(V.131) this yields

$$t(x) = \frac{2}{9}\alpha(\beta + 1)(\beta + 2)(\beta + 4)\ x^\beta, \qquad \text{(V.135)}$$

with α=2.01, β=-0.4. These results are shown in Fig.V.16.

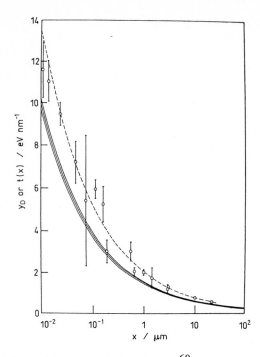

Fig.V.16 A proximity function for ^{60}Co (solid line) has been calculated from experimental data for y_D using Eq(V.131). The dashed curve is a fit to the y_D data. Also shown is an estimated error band on the calculated proximity function (after Zaider et al, 1982).

V.4.3.2 Proximity Functions for Diffused Charged Particle Tracks

In this section we discuss the following problem: Assume charged particle tracks characterized by a proximity function t(r). Assume further that each transfer point is affected by a diffusional process, i.e. is displaced randomly in space according to a Gaussian probability distribution function centered at the original position of the point, and with one-dimensional standard deviation σ. Consider now the three-dimensional Gauss transform:

$$G(x, y, z) = \frac{1}{\pi^{3/2} \beta^3} \int \int \int \exp\left[-\frac{(x - x')^2 + (y - y')^2 + (z - z')^2}{\beta^2}\right]$$

$$\times F(x', y', z')d'xd'yd'z.$$

$$(V.136)$$

It is easy to show that if $F \geq 0$ and $\iiint F(x,y,z)dxdydz=1$, then G is also non-negative and normalized to unity. Thus F and G can be regarded as probability distributions. If F is spherically symmetric

$$F(x, y, z) = f\left[(x^2 + y^2 + z^2)^{1/2}\right],$$
(V.137)

then G will have the same property. The integral can then be simplified by choosing a system of axes such that

$$(x, y, z) = (0,0, r)$$
$$(x', y', z') = (r' \sin \theta \cos \phi, r' \sin \theta \sin \phi, r' \cos \theta).$$
(V.138)

Thus:

$$g(r) = \frac{2}{\sqrt{\pi} \beta^3} \int_0^\infty \int_0^\pi \exp\left[-\frac{1}{\beta^2}(r^2 + r'^2 - 2r'r \cos \theta)\right] \sin \theta \, d\theta \, f(r') \, r'^2 \, dr'$$

$$= \frac{1}{\sqrt{\pi} \beta r} \int_0^\infty \left[\exp\left(-\frac{(r - r')^2}{\beta^2}\right) - \exp\left(-\frac{(r + r')^2}{\beta^2}\right)\right] f(r') r' dr'.$$

(V.139)

One may now identify $r^2 g(r)$ with $t(r)/T$ where

$$T = \int_0^\infty t(r) dr$$
(V.140)

and $r^2 f(r)$ with $t(r,\sigma)/T$ - the diffused proximity function ($\beta = 2\sigma$). With this one may rewrite Eq(V.139) in the form proposed by Chmelevsky et al (1980):

$$\frac{t(r, \sigma)}{r} = \frac{1}{2\sqrt{\pi} \sigma} \int_0^\infty \left\{\exp\left[\frac{-(u - r)^2}{4\sigma^2}\right] - \exp\left[\frac{-(u + r)^2}{4\sigma^2}\right]\right\} \frac{t(u)}{u} du.$$

(V.141)

The normalization of F implies that

$$\int_0^\infty t(r) dr = \int_0^\infty t(r, \sigma) dr = T$$
(V.142)

The statement (Kellerer and Rossi, 1978) that $t(r,\sigma)/r^2$ is monotonically decreasing results from the observation that the probability of finding an energy transfer point at distance r from the original position decreases with r.

V.4.3.3 Proximity Functions obtained from Cloud-Chamber Data

The application of cloud chambers in experimental microdosimetry has been discussed in Chapter IV. It is clear that, once the coordinates of energy transfer points (here ionizations) are known, one may obtain in a straightforward manner corresponding proximity functions of energy transfer. Although the actual energy deposited locally in an ionization event is not obtainable from a cloud chamber measurement, it is a reasonable approximation to associate with each ionization the average energy (W value) spent for producing an electron-ion pair in the gas.

During the condensation process droplets may diffuse away from the original site of the ionization event. This diffusion, which is of the order of 6-7 nm when scaled to the density of liquid water ($1g/cm^3$) is insignificant relative to conventional microdosimetry (that is volumes with dimensions of the order of micrometers) but it may distort quite seriously the results for spectra in nanometer-sized sites. In the radiobiological applications of microdosimetry it is in fact these latter (i.e. nanodosimetry) that are relevant and to this extent cloud chambers may be of only limited usefulness. To further appreciate this point we present in Table V.2 calculated values for the dose-averaged mean energy, e_D, deposited by 1.5-keV electrons in spherical sites ranging from 5 to 500 nm. In the column under $e_D{}^{(a)}$ the exact values of this quantity are given. The last column ($e_D{}^{(c)}$) shows results obtained from tracks distorted by typical cloud-chamber diffusion ($\sigma=4$ nm, see below). It is clear that even in sites as large as 100 nm significant differences remain if one were to approximate a track with its measured diffused counterpart.

Generally, it is not possible to "remove" the effects of diffusion from microdosimetric spectra. However, in the important case of proximity functions mathematical procedures have been developed for obtaining non-diffused functions (Zaider and Minerbo, 1988a and b). These are described below.

A. Fourier deconvolution

For convenience we rewrite Eq(V.141) that relates a proximity function, t(x), to its "diffuse" counterpart, $t(x;\sigma)$:

$$\frac{t(x;\sigma)}{x} = \frac{1}{2\sigma\sqrt{\pi}} \int_0^\infty \left[e^{-(u-x)/4\sigma^2} - e^{-(u+x)/4\sigma^2} \right] \frac{t(u)}{u} du. \qquad (V.143)$$

This expression is a Fredholm integral equation of the first kind to be solved for t(u).

Consider an analytic continuation of t(x) (which is normally defined only for positive values of the argument) as follows:

$$t(-x) = t(x) \qquad (V.144)$$

With this Eq(V.143) becomes:

$$\frac{t(x, \sigma)}{x} = \frac{1}{2\sigma\sqrt{\pi}} \int_{-\infty}^{+\infty} e^{-(x-u)^2/4\sigma^2} \frac{t(u)}{u} du. \qquad (V.145)$$

Let now $T(s,\sigma)$, $T(s)$ and $N(s)$ be the Fourier transforms of $t(x,\sigma)/x$, $t(x)/x$ and $\exp(-x^2/4\sigma^2)$, respectively. In Eq(V.145) $t(x,\sigma)/x$ is the convolution of a Gaussian distribution and $t(u)/u$. According to the well known convolution theorem (Bracewell, 1978):

$$T(s, \sigma) = \frac{1}{2\sigma\sqrt{\pi}} N(s)T(s) \qquad (V.146)$$

$N(s)$ may be calculated by direct integration:

$$N(s) = 2\sigma\sqrt{\pi} \, e^{-4\pi^2\sigma^2 s^2}. \qquad (V.147)$$

By introducing Eq(V.147) in Eq(V.146) one obtains:

$$T(s) = e^{+4\pi^2\sigma^2 s^2} \, T(s, \sigma). \qquad (V.148)$$

One may now revert to coordinate space (x) by applying the Fourier transform to Eq(V.148). The result, and solution of Eq(V.143), is:

$$\frac{t(x)}{x} = \int_{-\infty}^{+\infty} e^{i2\pi xs+4\pi^2\sigma^2 s^2} ds \int_{-\infty}^{+\infty} \frac{t(u, \sigma)}{u} e^{-i2\pi us} du \qquad (V.149)$$

The expression, Eq(V.149), suggests the following procedure for obtaining $t(x)$: a) find the Fourier transform (FT) of the diffused function $t(u,\sigma)/u$, b) apply the exponential filter $\exp(+4p^2\sigma^2 s^2)$ which enhances the high frequencies[22], and c) take the inverse Fourier transform (IFT) of the result.

The actual implementation of this method makes use of so-called discrete Fourier transforms for which efficient computer algorithms have been developed. In the following a few numerical examples are given.

Fig.V.17 shows calculated proximity functions for photoelectrons generated by Al soft x rays (the electron energies are 0.5 and 1.0 keV). Also shown in this figure

---

[22] s is measured in units of inverse wavelength (cycles/m); however, for convenience, we shall continue to call s a "frequency".

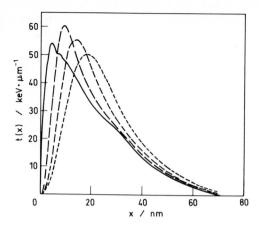

Fig.V.17 Calculated proximity functions for photoelectrons generated by Al soft x rays. The dashed curves show the effect of diffusion on the proximity functions; from left to right they correspond to σ= 3, 4 and 5 nm, see Eq(V.143) (after Zaider and Minerbo, 1988a).

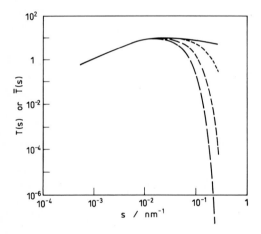

Fig.V.18 The Fourier transforms, T(s,σ) of the proximity functions of Fig.V.17. From top to bottom σ = 0, 3, 4 and 5 nm (after Zaider and Minerbo, 1988a).

are diffused proximity functions, t(x,σ), with σ=3, 4 and 5 nm. Fig.V.18 shows the Fourier transforms, T(s,σ), for the same values of σ. As expected, the general effect of diffusion is to narrow down the range of frequencies of the Fourier transform[23]. In the actual calculation of T(s,σ) the original function, t(x,σ)/x, is

---

[23] The basis of this observation is the fact that to a constant function, $h(t)=k$, there corresponds in the frequency space a Dirac delta function, $H(f)=k\delta(f)$, i.e. a single frequency; and conversely, to $h(t)=k\delta(t)$ corresponds $H(f)=k$.

sampled (that is, assumed known) at a finite number N of x points. For simplicity, the points are taken here as equidistant:

$$x_k = \Delta k, \quad k = 0,1,\ldots, N - 1 \qquad (V.150)$$

Here Δ is the sampling interval. The curves in Fig.V.18 were calculated by applying Eq(V.148) in its discretized version:

$$T\left(\frac{n}{\Delta N}, \sigma\right) = \exp\left[4\pi^2\left(\frac{\sigma}{\Delta}\right)^2\left(\frac{n}{N}\right)^2\right] T\left(\frac{n}{\Delta N}, \sigma\right) \qquad (V.151)$$

Because of the exponential factor in this equation, for certain values of σ, Δ and N, the function $T(s,\sigma)$ will span an extremely large range of values. For instance, in the calculations of Fig.V.18 $\Delta=0.5$ nm, $\sigma/\Delta=20$ and $N=8192$; then $T(s,\sigma)$ will change by about 1600 orders of magnitude over the full frequency range. Since no data (or computer) can claim this kind of accuracy, when calculating $T(s,\sigma)$ one needs to limit the frequency band to the range of meaningful values of $T(s,\sigma)$. As an example, a double-precision calculation on a Cray X-MP computer will handle $T(s,\sigma)$ values down to about 10^{-30} only (compare this with 10^{-1600} !). The procedure of ignoring frequencies above a certain value, n_c, is known as "applying a rectangular frequency window"; formally a filter of the form:

$$w(n) = \begin{cases} 1 & n < n_c \\ 0 & otherwise \end{cases} \qquad (V.152)$$

Figure V.19 shows the result of unfolding the diffusion from $t(x,\sigma)$ for $\sigma=4$ nm. The "ringing" in the unfolded curve is the direct result of using the rectangular frequency window. If this ringing is averaged over individual sine waves the resulting points (as shown in Fig.V.19) indicate essentially the spatial resolution with which $t(x,\sigma)$ can be unfolded[24]. With this caveat, the agreement between the diffusion-free points and the exact proximity function, $t(x)$, is excellent.

The ringing effect due to the particular rectangular window used in Fig.V.19 can be significantly decreased by using better filters. For instance, one may use a

---

[24] As a general rule, the larger the value of σ (or, similarly, the noisier the data $t(x,\sigma)$) the narrower the utilizable frequency range and therefore - by Nyquist theorem (Bracewell, 1978) - the poorer the resolution in the reconstructed $t(x)$.

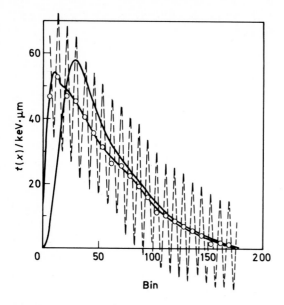

Fig.V.19 When the diffusion is "unfolded" from the proximity function t(x,σ =4 nm) with the aid of a rectangular frequency filter, see Eq(V.152), one obtains an un-diffused proximity function that shows a ringing pattern. However, when averaged over single wave length one obtains results (circles) in very good agreement with the (here known) original proximity function. The two solid curves show t(x, σ =0) and t(x, σ =4 nm). After Zaider and Minerbo (1988a).

Gaussian filter:

$$w(n) = \exp\left[-\frac{1}{2}\left(\frac{\alpha n}{n_c}\right)^2 \right] \tag{V.153}$$

where a is an adjustable parameter and n_c is the upper limit of the frequency band.

Noise associated with the finite computer precision (and the resulting need for frequency filtration) is only one consideration in evaluating the accuracy of the Fourier transform based unfolding algorithm. Another - and perhaps more critical element -is the experimental noise (uncertainty) in the cloud chamber data. It will be assumed here that this uncertainty is of statistical nature only, that is, due to the finite number of tracks recorded and/or analyzed. Published cloud chamber data indicate a spatial resolution of the proximity function of about 0.5 nm (when scaled to the density of liquid water, 1 g/cm^3). Uncertainties in the proximity function data appear to be, on the average, of the order of 10%. How critical are these uncertainties ? The answer clearly depends on a) the quantity to be calculated with the aid of t(x), b) the sensitivity of this quantity to the details of the shape of t(x), and c) the overall accuracy required. The e_D data in Table V.2 may be used for illustrating these points. As already mentioned at the beginning of this section this quantity is often used as an index of biological effectiveness. This quantity is obtained as follows [see Eqs(V.22,101)]:

$$e_D (d) = \int_0^\infty t(x)\left[1 - 1.5(x \ / \ d) + 0.5(x \ / \ d)^3\right] dx. \qquad (V.154)$$

Table V.2

d(nm)	$e_D^{(a)}$	$e_D^{(b)}$	$e_D^{(c)}$
5	15	21	0.4
10	82	94	5.0
20	260	264	51.0
50	803	800	468.0
100	1530	1524	1264.0
500	3154	3147	3092.0

Table V.2 contains values of e_D calculated with the exact function $t(x)$ and with the unfolded function. The three values (a,b,c) correspond, respectively, to the exact, unfolded and diffused $t(x)$. The agreement between $e_D^{(a)}$ and $e_D^{(b)}$ is satisfactory.

A different and more powerful approach for deconvoluting diffusion from proximity functions obtained from cloud-chamber data is presented below; this procedure, which is simpler and relatively insensitive to the statistical uncertainties of the data, is based on maximum entropy (MAXENT) and Bayesian methods (Appendix V.6).

B. MAXENT deconvolution

In order to apply MAXENT consider a discretised form of Eq(V.143):

$$t(x_i , \sigma) = \sum_{j=1}^J K(x_i , u_j)t(u_j), \quad i = 1,2,\ldots, I \qquad (V.155)$$

where the kernel K has an obvious definition, see Eq(V.143). Given I data points, $t(x_i,\sigma)$ and the matrix $K_{ij}=K(x_i,u_j)$ one looks for a solution, $t_j=t(u_j)$. The solution must satisfy two *a priori* conditions: a) non-negativity ($t_j \geq 0$), and b) normalization:

$$T = \sum_{j=1}^J t_j \qquad (V.156)$$

where T is the total energy in the track. From Eq(V.143) one may verify that:

$$T = \int_0^\infty t(x, \sigma)dx = \int_0^\infty t(u)du. \qquad (V.157)$$

With these restrictions (and possibly J>I) an exact or unique solution of Eq(V.155) may generally not exist. An additional complication is the fact that the data points are affected by experimental errors and Eq(V.155) becomes:

$$t_i(\sigma) = \sum_{j=1}^{J} K_{ij} t_j + d_i, \quad i = 1,2,\ldots,I \qquad (V.158)$$

where the uncertainties (noise terms) may be assumed to be normally and independently distributed: $d_i \sim N(0, \varepsilon_i)$. A solution to this problem may be obtained with the aid of Bayes's theorem (Jaynes, 1985), namely:

$$P(t_j \mid t_i(\sigma), I_0) = \frac{P(t_j \mid I_0) P(t_i(\sigma) \mid t_j, I_0)}{P(t_i(\sigma) \mid I_0)}. \qquad (V.159)$$

Here $P(A|B)$ means the probability of the stochastic event A given the prior event B. In words, Eq(V.159) states that the posterior probability of $\{t_j\}$, given the *data* $\{t_i(\sigma)\}$ and some other (unspecified for the moment) information I_0, can be calculated by multiplying the prior probability of $\{t_j\}$ given I_0 only, and the likelihood of obtaining the measured values $\{t_i(\sigma)\}$ if certain values $\{t_j(\sigma)\}$ and I_0 are assumed. The likelihood function in Eq(V.159) is:

$$Q = \exp\left[-\frac{1}{2} \sum_{i=1}^{I} \frac{\left(t_i(\sigma) - \sum_{j=1}^{J} K_{ij} t_j \right)^2}{\varepsilon^2_i} \right]. \qquad (V.160)$$

As regards the prior probability of $\{t_j\}$ one invokes the maximum entropy principle according to which:

$$P(t_j) = \exp\left[-\lambda \sum_{j=1}^{J} \frac{t_j}{T} \log \frac{t_j}{T} \right] \qquad (V.161)$$

and one should take for the set $\{t_j\}$ those values that maximize the prior distribution. The quantity in square brackets has been interpreted by Shannon

(1948) to represent the informational content of the "probabilities" $\{t_j/T\}$ and termed the entropy (H) of this distribution (see Appendix V.6). Maximizing the prior probability in Eq(V.161) is equivalent with maximizing the entropy H. Thus according to Jaynes (1957) by maximizing H one obtains the most non-committal (least biased) estimation of the solution $\{t_j\}$.

The photoelectron Al soft x ray proximity functions, analyzed above with Fourier deconvolution, have been also unfolded with the MAXENT method [for further details see Zaider and Minerbo, (1988b)]. These functions are shown in Fig.V.20 (full curves) together with the MAXENT unfolding result (broken curve), which is in very good agreement with the expected distribution. As before, to verify the sensitivity of this unfolding method to experimental noise in the data $\{t_i(\sigma)\}$ each value was randomly displaced according to a Gaussian distribution. For the two cases examined (relative noise 1% and 10%, $\sigma=4$ nm) no significant deterioration in the quality of the unfolded functions was observed.

The two conditions mentioned above (non-negativity and normalization) are satisfied automatically in MAXENT. Another important condition comes from the following result stated (without demonstration) by Kellerer and Rossi (1978): for any proximity function, $t(x)$, the quantity $t(x)/x^2$ must be a monotonically decreasing function of x. This condition (demonstrated in Section V.4.3.2) may be

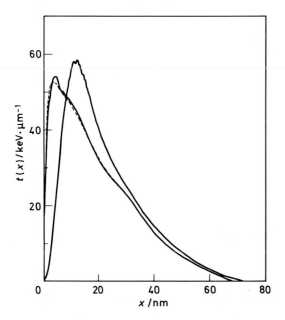

Fig.V.20 Proximity functions for photoelectrons generated by Al soft x rays: non-diffused (full curve, left) and diffused (full curve, right; $\sigma= 4$ nm). The dashed curve shows the result of MAXENT unfolding. After Zaider and Minerbo (1988b).

used to impose additional constraints on the solution t(x). An example is given in Zaider and Minerbo (1988b).

V.5 The Informational Content of the Moments of the Microdosimetric Distributions

Microdosimetric distributions are frequently used as input to biological models of radiation action, or for the purpose of estimating radiation protection quantities, for instance the dose equivalent. In this latter instance (this is discussed in more detail in Section VI.3) one may use an expression such as:

$$Q = \frac{1}{D} \int_0^\infty q(y) y f(y) dy \qquad (V.162)$$

to determine the quality factor, Q, for a radiation field characterized by the microdosimetric spectrum f(y). The function q(y) describes the biological effectiveness (per unit dose) of an increment y in lineal energy. For applications such as this to be practical it is important that f(y) be a *measurable* quantity. To the extent that microdosimetric volumes of the order of micrometer are utilized, proportional counters may be used, as explained in Chapter IV. There are however limitations related to the maximum pulse rate that can be achieved without appreciable pile-up (this is usually limited to 1000 events/s). At an absorbed dose rate of 10 mGy/h, which is typical for health physics measurements, a 10-cm diameter counter will register some 300 counts/s for low LET radiation but perhaps only 6 counts/s for fission neutrons. Since several thousands counts are necessary for evaluating f(y) one may need larger-diameter counters; this in turn will result in excessive pile up for the low-LET measurements. Another example of experimental limitation is the case of pulsed radiation sources where the instantaneous counting rate may very high and therefore force the health physicist to use a smaller counter. There are also theoretical and experimental indications suggesting that volumes of the order of 50 nm or less in diameter represent important radiation-sensitive cellular structures; yet, experimental microdosimetry is not available in this region.

Some of these difficulties may be obviated if one could use multi-event microdosimetric distributions, f(z;D). The measurement of these distributions consists of repeated measurements of the (multi event) deposition of specific energy, z, at equal absorbed doses, D. Since the magnitude of z can be controlled by the value of D (in practice, the duration of charge collection) it follows that one may employ ionization chambers with little or no gas amplification and thus determine f(z;D) in smaller simulated volumes. For similar reasons, multi-event measurements may be performed at very high dose rates. With f(z;D) thus obtained one may - at least in principle, see Eq(II.13) - recover the single-event distribution, $f_1(z)$ or f(y). Regrettably, it turns out that the numerical accuracy necessary for calculating $f_1(z)$ from f(z;D) is often beyond not only experimental capabilities but computer precision as well. This is illustrated in the following theoretical simulation (Zaider and Rossi, 1989): Consider an arbitrary microdosimetric spectrum, $f_1(z)$, and a sequence of multi-event spectra, f(z;D), obtained from f(z) with the aid of Eq(II.6):

$$f(z; D) = \sum_{k=0}^{\infty} e^{-n} \frac{n^k}{k!} f_k z(z), \qquad (V.163)$$

Table V.3 Comparison of the momenta of f(z) calculated directly, m_k, or by using Eq(II.20), m_k' (Zaider and Rossi, 1989).

k	m_k	M_k (n=96)	m_k
1	2.604	2.497	2.601
2	0.002511	6.475	0.002504
3	0.000724	17.44	0.000706
4	0.000320	48.75	0.000336
5	0.000170	141.5	0.000111
6	0.000101	425.9	0.000188
7	0.000065	1330.0	-0.000155

where n is the average number of events at dose D, $n=D/m_1$, and m_k is the k-th moment of $f_1(z)$. The actual calculation was performed with the computer code KFOLD developed by Kellerer (1985).The moments, m_k, can be calculated exactly. We have calculated the moments M_k of f(z;D) and then, with the aid of Eq(II.20), estimated the accuracy of the "unfolded" moments, m_k', obtained by solving this system of linear equations. The results, shown in the Table V.3, indicate that already the 5-th moment obtained in this way is in error by more than 50%, and increasingly erroneous results are obtained for higher-order moments. The source of these errors, ultimately responsible for the inapplicability of Eq(II.13), rests with the fact that the "tails" of the multi-events distributions, f(z;D), can not be known accurately.

Assume therefore that only a finite number of moments are known. By inquiring about the informational content of these moments we shall attempt to assess the extent to which microdosimetric derived quantities such as Q may be obtained from these moments only. The treatment of this problem is given below. For the sake of specificity we are using the important example of Q as defined in Eq(V.162). For the function q(y) we take the following expression (ICRU, 1986):

$$q(y) = \frac{a_1}{y}\left[1 - e^{-a_2 y^2 - a_3 y^3}\right] \tag{V.164}$$

$[a_1=5510 \text{ keV}/\mu\text{m}, a_2=5\times10^{-5} (\mu\text{m}/\text{keV})^2, a_3=2\times10^{-7} (\mu\text{m}/\text{keV})^3]$.

The formal statement of the problem is as follows: If the first K moments of a microdosimetric distribution, $f(y)$:

$$m_j = \int_0^\infty y^j\, f(y)dy, \quad j = 0,1,2,\ldots K \tag{V.165}$$

are known, find an unbiased estimate of $f(y)$, denoted $f_u(y)$, and determine the minimum number of moments, K, that brings a given functional, $G[f_u(y)]$, within a given interval (e.g. $\pm15\%$) of its true value, $G[f(y)]$. A general definition of G may be:

$$G[f(y)] = \int_0^\infty g(y)f(y)dy, \tag{V.166}$$

where g is known. Compare this expression with the definition of Q in Eq(V.162). The problem of estimating a distribution function from its moments is well known in statistics. The approach presented here is based on the maximum entropy method (MAXENT, see Appendix V.6) as applied to the following version of Eq(V.165) where only discrete values of y are considered:

$$m_j = \sum_{i=1}^n y^j_{\ j}\, f_i, \quad j = 0,1,\ldots, K \tag{V.167}$$

Here $f_i=f(y_i)$. The solution to this problem is (see Appendix V.6):

$$f_i = \frac{\exp\left[-\sum_{j=0}^K \lambda_j\, y_i^{\ j}\right]}{Z(\lambda_0, \lambda_1, \ldots, \lambda_K)} \tag{V.168}$$

where Z is the partition function:

$$Z(\lambda_0, \lambda_1, \ldots, \lambda_K) = \sum_{i=1}^n \exp\left[-\sum_{j=0}^K \lambda_j\, y_i^{\ j}\right] \tag{V.169}$$

The constants λ may be obtained (see Appendix V.6) by solving the following system of non-linear equations (efficient numerical algorithms are available for this purpose):

$$m_j = -\frac{\partial}{\partial \lambda_j} Z(\lambda_0, \lambda_1, \ldots, \lambda_K), \quad j = 0,1, \ldots, K \qquad \text{(V.170)}$$

or equivalently:

$$\sum_{i=1}^{n} y_i^{j} \prod_{j=0}^{K} e^{-\lambda_j y_i^{j}} - m_j = 0, \quad j = 0,1, \ldots, K \qquad \text{(V.171)}$$

Table V.4 compares values of Q for these four radiations obtained with exact distributions, f(y), or the MAXENT solutions, $f_u(y)$.

Table V.4   The quantity Q calculated using the exact and the estimated microdosimetric spectra.

Radiation	Exact Q	Estimated Q	
		K=2	K=3
15-MeV e	0.51	0.51	0.51
250-kVp x	1.1	1.1	1.1
14-MeV n	6.8	7.4	7.2
^{239}Pu α	26.0	26.4	26.3

The agreement is good and certainly within the accuracy requirements of Q in health physics applications. The analysis illustrated in Table V.4 was performed by assuming that the moments, m_j, are known exactly. This is of course hardly the case for moments determined experimentally, and any uncertainty thus introduced will reflect in the estimated values of Q.

The measurement of m_1 and m_2 requires a determination of n, the mean number of events at dose D (m_1 is equal to D/n; the ratio m_2/m_1 may be determined by the variance method).

V.6 Appendix

The Maximum Entropy Principle

In several sections of this book the so-called maximum entropy principle (MAXENT) is invoked. Here we attempt a systematic and elementary presentation of this principle.

Originally MAXENT has been introduced by Jaynes (1957) in a paper entitled "Information theory and statistical mechanics". Quite generally, Jaynes examines in this paper the following problem: Consider a random variable, x, that takes discrete values, x_1, x_2,...,x_n, with probability $p_1,p_2,...,p_n$, respectively. The probabilities p_i are not known. The only information we have in this respect is the normalization condition:

$$\sum_{i=1}^{n} p_i = 1, \qquad (V.172)$$

together with r "average" values of known functions, $f_j(x)$, j=1,...,r:

$$\phi_j = \sum_{i=1}^{n} p_i \, f_j(x_i). \qquad (V.173)$$

For instance, if $f_j(x)=x^j$ then we know the first r moments of x. The question posed is to infer the "best" (in the sense of least biased) estimate of the probabilities p_i.

To answer this question Jaynes suggests to use the entropy of the probability distribution, p_i, as a measure of its amount of uncertainty as follows: According to Shannon (1948) for any distribution probability the quantity:

$$H(p_1, p_2, \ldots, p_n) = -k \sum_{i=1}^{n} p_i \log(p_i) \qquad (V.174)$$

satisfies uniquely the conditions necessary for a measure of the amount of uncertainty: a) it is positively defined (k is an arbitrary positive constant), b) it increases with increasing uncertainty, and c) it is additive for the combination of any independent probability distributions. As an example, if there is *no uncertainty* as to the value of x (say, $p_1=1$ and all others are zero) then H=0. At the other extreme, maximum uncertainty obtains when:

$$\frac{\partial H}{\partial p_i} = 0, \quad i = 1,2,\ldots,n \qquad (V.175)$$

which is satisfied when all p_i are equal, as expected. The MAXENT principle asserts then that the least biased solution $\{p_i\}$ is that which maximizes H subject to the conditions, Eqs(V.173). Indeed, as Jaynes points out, any other solution (corresponding to a smaller value of H) has less uncertainty and therefore implies some additional knowledge on $\{p_i\}$ which we do not have [or else we would have use it as part of the conditions, Eqs(V.173)]. MAXENT can be now applied with the aid of the Lagrange multiplier technique. This is briefly reviewed below.

First, let us rewrite the conditions, Eqs(V.172,173), as follows:

$$g_1(p_1, p_2, \ldots) = \sum_{i=1}^{n} p_i - 1 = 0$$

$$g_2(p_1, p_2, \ldots) = \phi_1 - \sum_{i=1}^{n} p_i f_1(x_1) = 0 \qquad \text{(V.176)}$$

$$\ldots$$

$$g_r(p_1, p_2, \ldots) = \phi_r - \sum_{i=1}^{n} p_i f_r(x_i) = 0$$

For each of the r+1 conditions, g_j, one introduces a Lagrange multiplier, λ_j. The Lagrange technique (which finds the maximum of H subject to the conditions, Eqs(V.172,173)) consists of solving the system of equations, Eq(V.176), together with:

$$\frac{\partial}{\partial p_i} \left[H - \sum_{j=1}^{r+1} \lambda_j g_j \right] = 0 \qquad \text{(V.177)}$$

simultaneously for the unknowns p_i (i=1,2,...,n) and λ_j (j=1,2,...,r+1). It can be immediately verified that the maximum entropy solution is:

$$p_i = \exp\left[-\lambda_0 - \sum_{j=1}^{r} \lambda_j f_j(x_i) \right] \qquad \text{(V.178)}$$

where the unknowns λ are obtained from:

$$\lambda_0 = \log(Z)$$

$$\phi_j = -\frac{\partial}{\partial \lambda_j} \log(Z), \quad j = 1, \ldots, r \qquad \text{(V.179)}$$

The quantity Z is termed the partition function of the problem and is defined as

$$Z(\lambda_1, \lambda_2, \ldots) = \sum_{i=1}^{n} e^{-\lambda_1 f_1(x_i) - \lambda_2 f_2(x_i) - \ldots - \lambda_r f_r(x_i)} \qquad \text{(V.180)}$$

In practical applications the quantities ϕ_j are known from experiment and are therefore subject to experimental errors. An extension of the MAXENT method to this situation has been given by Gull and Daniel (1978). Consider a modified version of the set of conditions, Eq(V.173).

$$\phi_j = \sum_{i=1}^{n} p_i f_i(x_i) + d_j, \quad j = 1,2,\ldots,r \qquad (V.181)$$

where the noise terms, d_j, are assumed - as is customary - to be normally and independently distributed: $d_j \propto N(0,s_j)$. To infer values for the unknown probabilities, p_i, we can apply Bayes' theorem and write:

$$P(p_i|\phi_j) = \frac{P(p_i)P(\phi_j|p_i)}{P(\phi_j)}. \qquad (V.182)$$

In this expression $P(A|B)$ means the conditional probability of the event A given the (prior) event B. Eq(V.182) states that the posterior probability of a given set $\{p_i\}$, given the data $\{\phi_j\}$, can be calculated as the product of $P(p_i)$ - the *prior* probability of $\{p_i\}$ - and the likelihood of obtaining the measured values $\{\phi_j\}$ assuming the given set $\{p_i\}$. The likelihood function in Eq(V.182) is:

$$Q = e^{-\frac{1}{2}\sum_{j=1}^{r}\left[\phi_j - \sum_{i=1}^{n} p_i f_j(x_i)\right]^2 / \sigma^2_j} = e^{-\chi^2}. \qquad (V.183)$$

As regards the prior probability one may, again, use the entropy H. Gull and Daniel (1978) have suggested the following justification for this choice: Assume each value p_i being an integer number, N_i, of some (arbitrarily small) quantity, Δ. If configurations $\{p_1, p_2, \ldots, p_n\}$ are created by placing randomly "quanta" Δ in n boxes, the probability of a certain set $\{p_i\}$ is given by the binomial distribution:

$$w = \frac{N!}{N_1! N_2! \ldots N_n!}, \quad N = \sum_{i=1}^{n} N_i. \qquad (V.184)$$

At the limit of very small Δ (i.e large numbers N_i) Eq(V.184) can be transformed using Sterling's approximation to:

$$w = e^{-N\sum_{i=1}^{n}\frac{N_i}{N}\log\left(\frac{N_i}{N}\right)} = e^{-\frac{1}{\Delta}\sum_{i=1}^{n} p_i \log(p_i)} \qquad (V.185)$$

In the exponent of this expression one recognizes the entropy of the probability distribution. The problem is thus reduced to evaluating the maximum of the functional:

$$L[\{p_i\}] = H - \frac{\Delta}{2} \chi^2 \qquad (V.186)$$

Δ is an unknown constant; it is typically estimated from the requirement that χ^2 be equal to the number of degrees of freedom.

Chapter VI
Applications of Microdosimetry in Biology

VI.1 Radiobiology

VI.1.1 Introduction

The interaction of ionizing radiation with matter is invariably followed by an involute chain of processes. Thus the chemical consequences of irradiation of as simple a substance as pure water are still not entirely understood. It might therefore appear to be futile to attempt to account for the observed effects on the vastly more complex biological organisms in terms of the pattern of absorption of radiant energy. The justification for useful activity in what has been termed *radiation biophysics* is that in all their intricacy biological processes can be governed by simple fundamental mechanisms. An outstanding example is Mendelian genetics in which the general rules of inheritance were identified. This preceded knowledge of the organization and indeed even of the existence of DNA.

Except for partial cell irradiation, a technique primarily developed by Zirkle (e.g. Zirkle, 1957) experimental radiobiology involves random charged particle traversal of cells and the numerical specification of effects is usually expressed in terms of the fraction of affected cells, tissues or whole organisms. This is a determination of the probabilities of the effects of absorbed doses and major applications of microdosimetry consist in attempts to correlate these probabilities with the probabilities of energy deposition in subcellular volumes in which the elementary damages, termed *lesions* are produced. In principle the aim of these efforts is to derive the geometric disposition of the radiation sensitive portions of cells, located in what is here termed the (*sensitive*) *matrix* and if possible to identify them as known cytological entities. In practice the target is generally assumed to be the DNA of the cell and the theoretical objective is restricted to the accounting of radiation effects in terms of energy absorption in known or assumed configurations of DNA. Considerations based on microdosimetry can involve not only the distribution of absorbed energy in space but also in time, i.e. the dependence of biological effectiveness on absorbed dose rate as well as on radiation quality. Occasionally the joint influence of other variables (e.g. oxygen tension) is considered.

The spatial and temporal distribution of absorbed energy is only one of the random factors determining radiation effect probability. Organisms vary in radiation sensitivity and even in monoclonal cells the sensitivity varies with stage in the mitotic cycle (Sinclair and Morton, 1964). Furthermore, cells of identical characteristics and containing the same pattern of absorbed energy may vary in response because of the stochastic nature of biological processes.

This chapter is concerned with the contribution of microdosimetry to the understanding of radiation-induced biological effects. Models of radiation action must satisfy several general criteria; they must be: a) consistent with the facts of microdosimetry, b) in agreement with radiobiological data, and c) as simple as possible.

The second criterion is quite straightforward; unfortunately, it is often taken as the *only* criterion by which models need to be judged. An example of the first criterion is the microdosimetric requirement that dose effect curves (measured or calculated) be always limited from above by a straight line that indicates the fraction of biological objects (e.g. cells) that experienced an energy deposition event. A precise mathematical definition of this test, as well as other similar criteria, is given in the next section. It is important to understand that these tests have to do with *logical consistency* and are independent of any presumed (or known) biological mechanism of radiation action. Such tests are termed, perhaps inappropriately, as being "model free".

The third criterion (simplicity) is best illustrated with an example from statistics. It is always the case that if a set of data may be fitted with, say, a quadratic polynomial, a higher order polynomial will provide an improved fit as measured, for instance, with the standard χ^2 test. However, statisticians have devised methods (e.g. the F test) that determine whether the improvement in the fit obtained with a higher-order polynomial is *statistically significant*. As a matter of convenience (and certainly elegance) one would then retain the *simplest* analytic expression that, within the limitations of these tests, satisfies the data.

VI.1.2 Microdosimetric Constraints on Biophysical Models

In this section we list several examples of constraints imposed by microdosimetry on biophysical models of radiation action. These constraints are independent of any assumed mechanism by which radiation acts on the biological system. The following considerations are restricted to cases where the biological response is entirely due to energy absorption in a circumscribed region. In particular, they concern cells that exhibit an *autonomous* response, i.e. a response that is unaffected by energy absorbed outside of the cell. Hence, the following does not apply to a complex effect such as carcinogenesis that can be initiated by transformations in individual cells but may be controlled by intercellular mechanisms. However, as will be shown later, certain arguments based on microdosimetry permit conclusions on the RBE for multicellular effects. For simplicity of language, we shall here take a cell to represent the biological system; these concepts apply however to any system.

We shall denote by E(D) the probability of *effect* (i.e. a given end point) following exposure to dose D; the complementary quantity is the *survival*[25] *probability*, S(D)=1-E(D). It will be noticed that in a particular cell the effect may have occurred after a dose D', smaller than D. If we take this *actual* dose, D' (an unknown, *stochastic* quantity) to be a random variable then:

$$E(D) = Prob[D' < D] \qquad \text{(VI.1)}$$

E(D) is the cumulative distribution function of D'. The differential probability distribution function (pdf) e(D)dD is defined as:

$$e(D)dD = Prob\{D' \in [D, D + dD]\} \qquad \text{(VI.2)}$$

The following relations follow:

$$E(D) = \int_0^D e(D')dD'$$
$$e(D) = \frac{dE(D)}{dD} = -\frac{dS(D)}{dD} \qquad \text{(VI.3)}$$

For completeness we also introduce the *hazard function*, $\lambda(D)$, defined as follows[26]:

$$\lambda(D)dD = Prob\{D' \in [D, D + dD] \mid D' > D\} \qquad \text{(VI.4)}$$

According to this definition $\lambda(D)dD$ is the probability that the effect occurred in the interval [D,D+dD] relative to those cells that remained unaffected by dose D (i.e. D'>D). The difference between e(D) and $\lambda(D)$ is best understood with the following hypothetical example: assume that 1000 cells are exposed to radiation. Let S(D) = 0.6 when D = 10 Gy and assume that the next dose increment of 1 Gy (so ΔD = 1 Gy) further inactivates 10 cells. Then, by definition,

---

[25] Terms such as " survival ", " inactivation ", " killing " or " lethality " are commonly employed for cells in place of a more precise criterion which is usually the inability of multiplication beyond a stated limit.

[26] In epidemiology one similarly defines the hazard function $\lambda(T)$, where T is survival time. $\lambda(T)$ is known as *force of mortality* or simply *mortality*.

$e(D)\Delta D = 10/1000 = 0.01$, while $\lambda(D)\Delta D = 10/(0.6\times1000) = 0.017$, that is $\lambda(D)$ refers to the fraction inactivated relative to those cells (600) still surviving after 10 Gy of radiation. The following relations obtain:

$$\lambda(D) = \frac{e(D)}{1 - E(D)} = \frac{e(D)}{S(D)}$$

$$\int_0^D \lambda(D')dD' = \int_0^D \frac{e(D')}{1 - E(D')} dD' = \tag{VI.5}$$

$$= - \log[1 - E(D)] = - \log S(D).$$

It is also evident that:

$$S(D) = e^{-\int_0^D \lambda(D')dD'}. \tag{VI.6}$$

The functions introduced so far, $e(D)$, $E(D)$, $S(D)$ and $\lambda(D)$, are equivalent to each other (i.e. they contain the same information) and selecting one or another is mainly a matter of convenience[27]. In the case where $e(D)$ does not exist [that is, $E(D)$ is discontinuous] the other functions continue to be well defined.

a) Dose-effect curves for autonomous cells must be *linear* at low doses.

In Chapter II we have shown that at a dose, D, that satisfies the condition $D/z_F<1$ the microdosimetric volume is traversed, on average, by less than one event. Let the *gross sensitive volume (GSV)* denote the smallest geometrical volume where energy absorption can cause the (biological) effect under consideration. The GSV clearly depends on the end point selected. For instance, for a cell one may take the volume of the cell nucleus to represent this quantity. If the single-event frequency-averaged specific energy, z_F, refers to energy deposition in the GSV then whenever $D/z_F\ll1$ an increase in dose, say a doubling of the dose, will simply double the number of cells traversed by events, and therefore double the effect; this is because cells continue to experience, on average, single events only. This demonstrates the linearity of the dose effect curve at low doses.

There are two important points to be made: Firstly, the criterion for "low dose" must be based on the number of *microdosimetric events*. Secondly, linearity refers to the dependency of effect on *dose*; as a function of, say, specific energy the effect $E(z)$ does not have to be linear. For instance, $E(z)$, might be a step function (i.e. there is a threshold in the amount of energy necessary to produce the effect) yet $E(D)$ will continue to be linear at low doses.

---

[27] The function $\lambda(D)$ has been termed *reactivity*, $R(D)$, by Hug and Kellerer (1966).

b) The dose effect curve for autonomous cells has a microdosimetric upper boundary.

If the radiation response of cells to radiation is *autonomous*, an obvious upper limit of effect probability is the probability that an (energy deposition) event occurred in it. The mean number of events at an absorbed dose D is Φ^*D where $\Phi^*=1/z_F$ is the *event frequency*. Since events are statistically uncorrelated Poisson statistics apply and the probability of no event is $\exp(-\Phi^*D)$. Thus

$$E(D) \leq 1 - e^{-D\Phi^*}$$

$$S(D) = 1 - E(D) \geq e^{-D\Phi^*}. \qquad (VI.7)$$

This is illustrated in Fig.VI.1. The inequalities in the expression, Eq(VI.7), are *local*, in the sense that they must be satisfied at each dose. An equivalent condition, in terms of the hazard function, is

$$\int_0^D \lambda(D')dD' \leq \frac{D}{z_F} = \frac{1}{z_F}\int_0^D dD'. \qquad (VI.8)$$

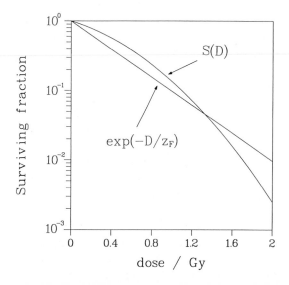

Fig.VI.1 Illustration of the constraints, Eq(VI.7): Any survival curve, S(D), must be limited from below by the curve representing $\exp(-D/z_F)$, that is by the fraction of cells that experience no energy deposition events. In this hypothetical situation the two curves intersect at a dose of approximately 1.4 Gy; for larger doses S(D) is incorrect because Eq(VI.7) is violated.

Certain analytic expressions frequently used in radiation biology clearly do not fulfill these conditions. For instance, for the linear-quadratic expression:

$$S(D) \;=\; e^{-\alpha D - \beta D^2} \tag{VI.9}$$

and

$$\lambda(D) = \alpha + 2\beta D \tag{VI.10}$$

but the hazard function can not be a continuously increasing function of dose without violating Eq(VI.7).

The quantity z_F depends on the size of the reference volume; for example, for spheres the larger the diameter the smaller z_F (see Fig.VI.2). Thus, if the size of the GSV is not known (as is usually the case) one may use the constraints, Eqs(VI.7,8), to estimate a *lower limit* for its dimensions which must apply if every event has unit probability of causing the effect. An interesting occurrence is when such a lower limit is actually *larger* than the physical dimensions of the irradiated object (e.g. cell); one would then conclude that cell-to-cell interaction contributes to the effect on single cells that are thus not autonomous.

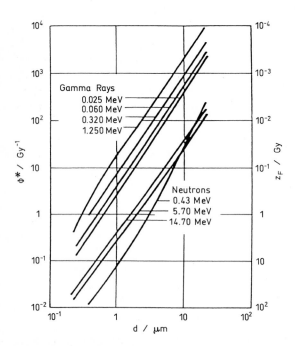

Fig.VI.2 Measured values of z_F, the dose-averaged specific energy, and its inverse, Φ^*, as a function of the site diameter. The data are for gamma radiation and for neutrons at the energies indicated. They exclude events due to delta rays only (after Rossi, 1979).

c) A non-linear dose-effect curve implies causation of the effect by multiple events and a *lower limit* for the average dimension of the site where these events must occur.

According to a) the initial portion of dose-effect curves is linear and represents causation of the effect by single events. Generally not every single event causes the effect. For instance, less than one in ten alpha particles that traverse the nucleus of certain cells in tissue culture causes lethality (Lloyd et al, 1979). This may be because some events do not produce lesions or because lesions have limited probability of causing the effect. In a survival curve, $S(D)=\exp(-\alpha D)$, $\alpha=f\Phi^*$ where f is the fraction of events causing the effect and Φ^* is the event frequency. If f is unknown one can identify a lower limit of the GSV which must be in any case large enough so $\Phi^*\geq\alpha$. This point has been made at b) above.

When the effectiveness of radiation varies non-linearly with D, i.e. when the hazard function is not constant, more than one event is involved in producing the effect. The effect is then, at least partly, due to the combined action of two or more events that result in radiation products (which need not be specified) that *jointly* cause the effect. Lesion formation requires that the *minimum range of distances* over which products may interact is at least as large as the diameter of a site within the GSV that contains two events at the absorbed dose, D. This statement is illustrated in the following.

The experimental data shown in Fig.VI.2 must be extrapolated to estimate the event frequency for ^{60}Co gamma radiation in very small volumes (of the order of nanometers). For d> 1 μm the slope equal to 2 in this logarithmic plot indicates that Φ^* is proportional to d^2. This is in accord with the fact that at in a given fluence the number of traversals of a site is proportional to the cross section. The data extend to 250 nm and show the onset of a steeper slope. This is due to traversal of sites without relevant energy transfer (one or more "ionizations"). When, in sufficiently small sites within an irradiated medium, the probability of energy transfer is small it is proportional to their volume (i.e. to d^3)[28].

An extrapolation of the curve in Fig.VI.2 that is based on $\Phi^*=Cd^2$ (with C=20 Gy^{-1} at d= 1 μm) thus involves a substantial overestimate of Φ^* at much smaller values of d. According to the applicable Poisson statistics, for a mean number of events equal to n the ratio between the probability of 2 events, p(2), and the probability of 1 event, p(1), is given by $p(2)/p(1)=n/2=\Phi^*D/2=Cd^2D/ = 10d^2D$. If d= 10 nm and D= 10 Gy this ratio is equal to 0.01. Hence at large doses the probability that a structure larger than, say, several DNA nucleotide pairs experiences two rather than one relevant event is substantially less than 1%. As

---

[28] For diameters that are large compared with the range of charged particles Φ^* is also proportional to d^3 because the event frequency is proportional to the number of tracks per unit volume.

shown below(section VI.1.2) single relevant energy transfers can disrupt the DNA molecule and the frequency for this process is *at most* equal to the number of such events. It may thus be concluded that in sites having diameters of 10 nm or less the biological effects at usually applied absorbed doses (several Grays) are almost exclusively due to single events with attendant linear dose-effect relations and independence of dose protraction. Conversely, *non-linear* dose-effect relations indicate that biological effects must be the result of multiple events in substantially larger sites.

d) The Kellerer-Hug theorem

A relation first stated by Kellerer and Hug (1972) sets up a *lower limit* for the *mean number of inactivating events*, and thus for the number of microdosimetric events at the *mean inactivation dose* (see below). This theorem expresses a *global* condition, that is it relates to the overall dose response curve.

The mean inactivation dose, Δ, is defined as follows:

$$\Delta = \int_0^\infty De(D)dD = \int_0^\infty S(D)dD. \qquad (VI.11)$$

The second relation obtains if one integrates by parts *and* if $S(D=\infty)=0$[29].

A second quantity necessary for this theorem is the variance of the inactivation dose:

$$\sigma^2 = \int_0^\infty (D - \Delta)^2 e(D)dD = 2 \int_0^\infty D\, S(D)dD - \Delta^2. \qquad (VI.12)$$

From the same considerations as those that lead to Eq(VI.7) it follows that $\Delta \geq z_F$.

Eqs(VI.11,12) may be expressed in terms of the *number of events* (a discrete quantity) rather than dose. At a given dose, D, the probability of exactly n events is:

$$P_v(D) = e^{-\frac{D}{z_F}} \frac{(D/z_F)^v}{v!} \qquad (VI.13)$$

Correspondingly, if E_v denotes the probability of effect in exactly v events (and also $S_v=1-E_v$), one has:

---

[29] In fact, as D→∞ S(D) needs to approach zero faster than 1/D.

$$E(D) = \sum_{v=0}^{\infty} E_v \, p_v(D) \qquad (VI.14)$$

The differential quantity, e_v, is the probability of effect by the v-th event:

$$e_v = E_v - E_{v-1} \qquad (VI.15)$$

The mean number of inactivating events, m, and the variance of v are:

$$m = \sum_{v=0}^{\infty} v \, e_v = \sum_{v=0}^{\infty} S_v,$$

$$\sigma_v^2 = \sum_{v=0}^{\infty} E_v (v-m)^2 = \sum_{v=0}^{\infty} (2v+1) S_v - m^2. \qquad (VI.16)$$

The link between the two sets of quantities, (Δ, σ) and (m, σ_v), is easily demonstrated:

$$\Delta = \int_{0}^{\infty} S(D)dD = \sum_{v=0}^{\infty} S_v \int_{0}^{\infty} p_v(D)dD = m z_F,$$

$$\sigma^2 = z_F^2 (\sigma_v^2 + m), \qquad (VI.17)$$

$$\frac{\sigma^2}{\Delta^2} = \frac{\sigma_v^2}{m^2} + \frac{1}{m}.$$

Note the analogy between the mean number of events, $n=D/z_F$, and the mean number of *inactivating* events, $m=\Delta/z_F$. Also, clearly $m \geq 1$ and therefore $\Delta \geq z_F$. It follows that:

$$\frac{1}{m} = \frac{\sigma^2}{\Delta^2} - \frac{\sigma_v^2}{m^2},$$

$$m \geq \frac{\Delta^2}{\sigma^2}. \qquad (VI.18)$$

The last statement is the mathematical expression for the *Kellerer-Hug theorem*. The quantity Δ^2/σ^2 has been termed by Hug and Kellerer (1966) *relative steepness*. According to this theorem the average number of microdosimetric events (which is necessarily larger than m) must be larger than the (experimentally determined) relative steepness. This automatically and in a model-independent manner sets up a lower limit for the dimensions of the GSV.

Because the average number of inactivating events, m, is given by Δ/z_F the Kellerer-Hug theorem may be also expressed as:

$$z_F \;\leq\; \frac{\sigma^2}{\Delta} \qquad\qquad (VI.19)$$

Compare this with the expression obtained above, Eq(VI.7):

$$z_F \;\leq\; \frac{D}{-\log[S(D)]}. \qquad\qquad (VI.20)$$

Finally, we mention without demonstration another limiting equation [Kellerer and Hug (1972), Eq(7.11)]:

$$z_D \;\leq\; \frac{\sigma^2}{\Delta}. \qquad\qquad (VI.21)$$

A plausible intuitive explanation of the theorem, Eq(VI.18), can be given as follows: Since events are by definition statistically independent they are subject to a Poisson distribution. When the mean number of microdosimetric events is n, the relative variance of the corresponding Poisson distribution is 1/n and the relative steepness is equal to n. Additional stochastic processes must increase the variance and consequently reduce the steepness. Hence, n cannot be less than the observed relative steepness. Since the absorbed dose is proportional to the number of events, the relative steepness of e(D) is, in the absence of the biological random factors, the same as that of the Poisson distribution about the mean number of events. The concept of relative steepness can be further illustrated by a few simple examples.

In the case of cell survival, the simplest mode of inactivation is by single events. This is represented by what has been termed a "*first-order*" or *one-hit* inactivation:

$$S(D) \;=\; e^{-\alpha D} \qquad\qquad (VI.22)$$

where S is the surviving fraction when the mean number of lethal events in cells caused per unit absorbed dose is equal to α. The *steepness*, s, of an exponential survival curve is 1. The steepness of measured curves, especially those obtained with low-LET radiation, can be considerably larger. Hug and Kellerer (1966) quote values up to 3.9 obtained in low-LET irradiation of various cell lines which implies a mean number of at least 3.9 interacting events.

In the "multi-hit" formulation it is assumed that cell killing is due to the occurrence of several events in a single target. If n events are required for cellular inactivation the survival curve is given by:

$$S(D) = \sum_{j=0}^{n-1} e^{-\lambda D} \frac{(\lambda D)^j}{j!}, \qquad (VI.23)$$

where λ is the number of effective hits per unit of absorbed dose. The relative steepness of this curve is n. This eliminates 2- or even 3-hit kinetics in the inactivation of most cell lines.

An alternate formulation is the multi-target model which refers to m targets that are damaged independently in single events. The survival is given by

$$S(D) = 1 - (1 - e^{-\lambda D})^m. \qquad (VI.24)$$

When m=2 the relative steepness is only 5/3.

If, in either of these models, there is a superposition in which n or m can be 1 as well as 2 (i.e. single hits or damage to single targets can also cause lethality) the relative steepness becomes even less.

Pairs of events are of primary importance in *dual radiation action* (see section VI.1.4). In this formulation

$$S(D) = e^{-\alpha D - \beta D^2}, \qquad (VI.25)$$

where $\sqrt{\beta}$ is proportional to the number of altered entities (sublesions) that are produced per unit absorbed dose and that combine to form lesions. When α is small the relative steepness is about 3.66 which is near the maximum cited by Hug and Kellerer. However, biological variability and linear component increase the difference.

It has been found nevertheless that the linear-quadratic expression for S(D) frequently is an adequate approximation to the observed survival curves and, except for the largest values cited by Hug and Kellerer, the underlying mechanism in which lesions are produced by interaction of pairs of sublesions leads to sufficiently large values of s. As shown above this is not the case if lesions are due to two events in one target or to single events in two targets. The latter of these models may seem to be equivalent to dual action but it does not incorporate a dependence on local concentration of damaged molecules.

It is of course possible to obtain larger values of S by increasing the number of hits or targets. Various models have also involved combinations in which several hits in several targets are postulated and these have been reviewed by Zimmer (1961). Other models making no reference to hits or targets rely on more general postulates. For example, the "saturation of repair" mechanism is based on the assumption that the damage that increases with absorbed dose is removed with only a limited repair capacity of the cell.   Without engaging in a critical examination of the validity of such alternate models it can be stated that they have not been based on microdosimetry.

VI.1.3 Empirical Data in Radiation Biology

1. Introduction

In this section we summarize some of the pertinent experimental findings in this field. This summary will be, by necessity, neither comprehensive nor exhaustive. Instead, emphasis is being placed on those conclusions that have found (or might find) explanatory elements in the field of microdosimetry.

The presentation here refers mostly to *general mechanisms* rather than to the precise (and unendingly complicated) details of actual radiobiological experiments. Numerical details, when invoked, are left for concrete examples illustrating the applicability of particular microdosimetry-based models of radiation action. For instance, following exposure to ionizing radiation modifications, generically termed *sublethal damage*, occur in cells. The nomenclature indicates that *singly* they cannot produce the effect (here, cellular lethality); however, an accumulation of sublethal damage may result in the reproductive inactivation of the cell. This process is controlled - among other things - by the fact that sublethal damage may be enzymatically repaired. As a consequence, one expects the outcome of radiation exposure to be influenced not only by dose, but by *dose rate* as well. The actual rate of sublethal damage repair may, of course, change depending on the cellular system (some cells simply lack a repair mechanism), irradiation conditions, etc. A microdosimetric description of the radiation field renders information on both the *spatial and temporal* distribution of sublesions, and these factors are expected to control the collective effect of sublesions. All these are "matters of principle", quite independent of their quantitative aspects. A mathematical formalism of radiation action may then be stated in terms of these factors (as usual, its validity is probed against available data).

2. Radiobiological effects and their geometric domains

2a. General considerations

Radiation biology is concerned with targets that range in size over at least 10 orders of magnitude between biomolecules and tissues of whole organisms. Most of the effects are ultimately due to DNA damage but in addition to their number the *spatial distribution* of the affected molecules is of cardinal importance in the causation of effects.

The utility of microdosimetry in the interpretation of biological effects in terms of radiation physics is generally restricted to cellular effects. The diameter of mammalian cells is of the order of 10 µm. Microdosimetry is not important in multicellular structures because at the absorbed doses of interest the number of events is so large that the specific energies differ negligibly from absorbed dose. The evaluation of small-scale variations of absorbed dose (e.g. around radioactive particles) which may be termed *minidosimetry*, is sometimes confused with microdosimetry. The difference is evidenced by the fact that in minidosimetry one is dealing with a non-stochastic (expectation) value of a point function that is subject to a continuous geometric distribution. A detailed treatment of the combination of microdosimetry and minidosimetry has been given by Roesch (1968).

The effects of radiation are measured in terms of two types of *end-points*: *structural changes* in targets, and *inactivation* of a certain biological function. Examples of the former are: single- and double-strand DNA breaks or chromosomal aberrations. Examples of the latter include: enzyme inactivation, mutation, cell death (inability to divide), cellular transformation or, at organism level, carcinogenesis.

The terms *non-stochastic* and *stochastic effects* have been applied to radiation effects. Stochastic effects are understood to be subject to a "yes-or-no" criterion. The scale of severity of stochastic effects is in terms of *probability* because changes in administered doses change only the *incidence* of such effects. Non-stochastic effects vary in degree and can, at least in principle, be specified by a numerical scale. Thus increasing doses to an organism cause increasing damage that may be specified on the basis of subjective levels of severity (such as "degree" of cataract of the lens of the eye) or in terms of a more objective scale that may, for instance, be based on physiological parameters such as blood count.

The distinction can be envisaged to be due to a semantic ambiguity of the term *effect*. Any degree of a non-stochastic effect is subject to a "yes-or-no" criterion of a stochastic effect. Hence the non-stochastic effect may be considered to be, in general, a collective of stochastic effects which (as in the example of blood count) may involve different cells. When the effect is always the same (e.g. the number of broken chromosomes and the probability of a chromosome break) a determination of the degree of stochastic effect (i.e. probability) on the basis of the degree of the non-stochastic effect (i.e. number) requires knowledge of the size of the irradiated population. At low probabilities the numerical specifications of effects by the two methods are proportional to each other.

It is often possible to distinguish in the cells of higher organisms (eukaryotes[30]) two greatly disparate regions where the absorption of radiant energy causes biological effects. This is of essential importance to interpretations of cellular radiobiology in terms of microdosimetry.

At the absorbed doses normally applied to biological systems, the mean number of events in the smaller *nanometer domains* is much less than 1 even for low-LET radiations. Consequently, first order inactivation dominates with effects being due to single events; the hazard function is constant and there is no dependence of effect probability on protraction of irradiation over periods in which the cellular sensitivity is constant.

In the larger *micrometer domain* multiple events may occur, especially for low-LET radiation and this is frequently of major importance in experimental radiobiology and radiotherapy. Enzymatic repair and any other restituting mechanism can alter the generally non-linear dose-effect curves.

2b. The nanometer domain

It is generally accepted that radiation-sensitive targets in the mammalian cell are contained in the cellular nucleus. In particular, the cellular effects of radiation are believed to be initiated by damage to the DNA molecule (for a review see Powell and McMillan, 1990)[31]. DNA (deoxyribonucleic acid) consists of two strands of nucleic acid linked to each other by means of hydrogen bonds. The basic unit in a polymer chain of nucleic acid is a *nucleotide*: it consists of a nitrogenous base [adenine (A), guanine (G), thymine (T) or cytosine (C)], a sugar (in the case of DNA: deoxyribose) and a phosphate group. The bases are further classified in purines (A and G) and pyrimidines (T and C). The deoxyriboses together with the phosphate links make up the backbone of the DNA strand. In cellular DNA the two strands are oriented such that the bases are paired (a purine always pairs with a pyrimidine): A-T or G-C.

In the Watson-Crick double-helical model of DNA (Fig.VI.3) the helix has a diameter of about 2 nm. A full turn of the helix is 3.4 nm in length and contains 10 nucleotide pairs; consecutive nucleotide pairs are rotated with respect to each other by 36^0. The typical "volume" occupied by a nucleotide pair is therefore 1 nm^3.

In a mammalian cell a DNA molecule contains of the order of 10^8 nucleotide pairs. The atomic mass of each pair of nucleotides is about 660. It follows that a human cell with 46 chromosomes contains about $5 \cdot 10^{-15}$ Kg of DNA material (by weight this is 6% of the cellular nucleus, assuming the whole nucleus, of diameter

---

[30] The term eucaryotic denotes cells that have a nucleus containing the cellular DNA.

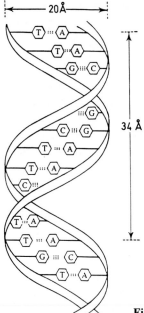

20 Å

34 Å

Fig.VI.3 The Watson-Crick model of DNA (after Baum, 1978).

5 mm, has the density of water). If the DNA in a human cell would be tightly packed it would occupy a volume of almost 4 μm^3.

The total length of cellular DNA (several cm) and the dimensions of the cellular nucleus (several μm) makes it clear that an extensive packaging of the DNA material is necessary. *Eucaryotic* DNA is tightly bound to proteins called *histones*. Several kinds of histones contribute to the formation of *chromatin*, a long structure with the appearance of "beads on a string". The beads (termed *nucleosomes*) are disc-shaped (diameter: 11 nm) and consist of the DNA double helix wound around a histone core. The "string" connecting the beads is called DNA *linker*. Some 60 nucleotide pairs (20 nm) separates a bead from the next one. The next level of packing consists of *chromatin fibers*; their diameter is 30 nm. Finally, prior to mitosis the fibers may condense in a series of loops that generate yet thicker structures (*chromosomes*) with diameters between 0.3 and 0.7 μm. The chromosomes display a series of visible bands, assumed to reflect a

---

[31] The minor role apparently played by radiation-induced protein damage has been attributed to the fact that cellular proteins (enzymes) are continuously replenished and therefore the inactivation of a moderate fraction of them may not be consequential. Quantitative studies on enzymes also indicate that doses of the order of 1000 Gy are necessary to effect 50% inactivation in the intracellular environment. This is several orders of magnitude larger than "conventional" radiobiological doses (several Gray or less).

micrometer-scale level of organization of the chromosome; however, the role of the bands remains unknown. Fig.VI.4 shows the different levels of organization of DNA material that lead to the cellular chromosome.

There is general agreement that radiation damage in cellular DNA occurs mainly in the form of single- (ssb) and double-strand breaks (dsb)[32]. The yield (number) of single-strand DNA breaks (ssb) changes *linearly* with dose for all radiations. For cellular DNA the yield of dsb depends also linearly on dose; however, for DNA in aqueous solution the dsb response is sometimes found to follow a linear-quadratic $(\alpha D + \beta D^2)$ curve. In dilute solutions DNA molecules can be affected by radicals that diffuse over appreciable distances while in the cellular environment the DNA is protected by the chromatin structure from the *indirect* effects of radiation. Thus, for most practical purposes, only nearly *direct* hits in the DNA duplex will contribute to the dsb yield.

Figure VI.5 shows a result, typical for a group of reports, for the variation of RBE with LET for DNA breaks generated intra-cellularly. The RBE for ssb induction shows a decreasing trend at higher LET values. In the case of dsb-s a peak is observed in the range of LET values of 100 to 200 keV/μm. The nearly constant yield below about 10 keV/μm is due to the fact that LET is a grossly inadequate quantity for the specification of energy deposition in nanometer sites. When such small volumes are traversed by particles of low or moderate LET even single collisions occur with an average frequency that is less than one. Hence an increase in LET does not change the energy deposited in an event. Fig.VI.6 shows the relation between y_D and LET for deuterons between 10 and 100 keV/μm (Rossi, 1994)[33]. Comparison with Fig.VI.5 shows that at low to moderate LET the yield of dsb-s parallels the dose-averaged lineal energy. At higher values of LET the yield declines because of saturation (i.e. absorption of more energy than necessary to produce the effect).

On the other hand, other researchers (Belli et al, 1991) appear unable to observe a variation of dsb RBE with LET. Evidence for the preponderance of direct effect in cellular DNA comes from a comparison of dsb induction in *dry* DNA (where only direct damage occurs) with data observed on viral DNA irradiated intra-cellularly: in both situations similar yields of dsb-s have been observed (Baverstock and Wills, 1989).

---

[32] Other types of damage are DNA-DNA cross links, DNA-protein cross links and base or sugar damage.

[33] The suggestion here that the yield of dsb-s depends on y^2 (the average value of which is related to y_D) corresponds to the mechanism where: a) two ssb-s in close proximity are necessary to produce one dsb, and b) ssb-s are produced at a rate proportional to y. These ideas (including the effects of saturation) have been further formalized by Zaider (1993).

DNA double helix 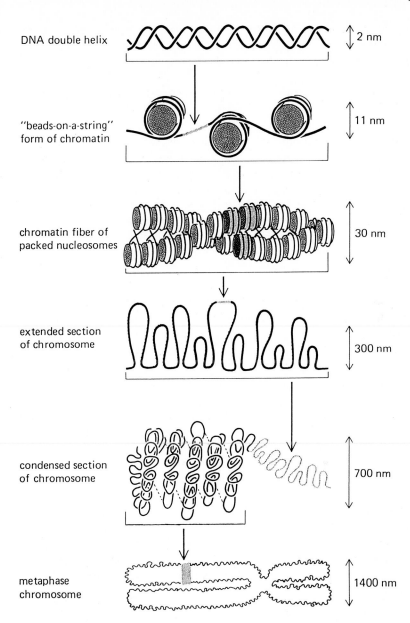 2 nm

"beads-on-a-string"
form of chromatin

11 nm

chromatin fiber of
packed nucleosomes

30 nm

extended section
of chromosome

300 nm

condensed section
of chromosome

700 nm

metaphase
chromosome

1400 nm

Fig.VI.4 Packing of DNA in increasingly larger structures leading to condensed chromosomes (source: Alberts et al, Molecular Biology of the Cell, 1983).

Typical numbers for the initial yield/Gy of DNA damage are: 40 dsb-s, 500-1000 ssb-s, 1000-2000 base or sugar alterations and 50-100 crosslinks. The observed (as opposed to initial) amount of DNA damage may, however, be different because of the possibility of repair.

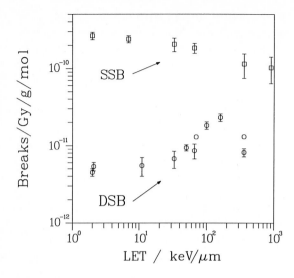

Fig.VI.5 The yield of single (top) and double (bottom) DNA strand breaks induced in V79 cells by radiations of different LET (Kampf, 1988).

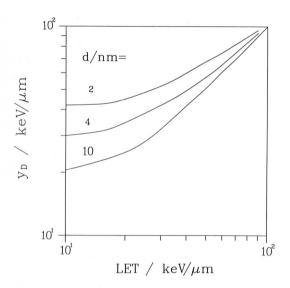

Fig.VI.6 Approximate relation between y_D and LET for deuterons. The site diameters are: 2, 4 and 10 nm.

Reports in the literature indicate that most ssb-s repair within about 10 min following irradiation. The situation for the rate of dsb repair is more ambiguous, but the half-time for repair is often longer, of the order of hours. A number of factors may affect the rate of dsb repair: a) the proximity of the two ssb-s responsible for the dsb ("blunt" ssb-s, opposite each other, are presumably more difficult to repair), b) the proximity of different dsb-s, c) the position of the dsb on DNA (transcribing regions of DNA repair more easily), and d) the position of the cell in the cell cycle. It is also important to note that what is generally termed "repair" involves two distinct steps: damage repair and recombination of the broken strands. This latter step may lead to *misrepair*.

Single interactions with DNA cause a variety of biological effects. The most immediate ones are simple *chromosome breaks* which can obviate cell proliferation because of loss of genetic information. *Mutations* can be due to direct damage of specific portions of DNA or, indirectly, result from the loss - in mitosis - of chromosome fragments on which they are located. Dose-response curves for mutation induction are generally linear-quadratic in dose for low-LET radiation and linear for higher-LET particles. As with other end points, mutations induced linear-quadratically as a function of dose show a sparing effect on lowering the *dose rate*; little, if any, such effect is observed for the case of linear response to radiation. As a function of LET the RBE for mutation induction in mammalian cells shows a curve similar to that obtained for dsb induction (Fig.VI.5) and cellular inactivation, namely with a peak at about 100-120 keV/μm. For low-LET radiations the mutation yield is 10^{-5}-10^{-6} per Gy, with a variation of the order of 10 across the LET spectrum.

It should be noted that while effects at the nanometer level result in linear dose-effect relations, the converse can not be assumed. Single events, particularly due to high LET particles, can cause multiple damage which results in the more complex mechanism operating at the micrometer level.

2c. The micrometer domain

The signature of the micrometer domain is the occurrence of effects that are not proportional to absorbed doses of less than tens of gray of low LET radiation. This is an indication of interaction between the products of relevant energy deposits that are separated initially by distances of the order of a micrometer or more. An important example are the so-called *two-hit* chromosome aberrations.

Chromosomal aberrations include single breaks and two-break aberrations[34]. Aberrations of both types are generally observed at metaphase at the first cell division. At the time of observation some of the chromosome or chromatin breaks may have repaired already[35]. One can further identify whether the break occurred

---

[34] A third category, changes in the chromosomal number, are particularly important when occurring in germ cells. The relation, if any, between the yield of these modifications and radiation is not clear.

[35] A new technique, known as premature chromosome condensation (PCC) allows early detection of chromosome or chromatin breaks before any significant repair took place.

before or after DNA replication: for instance, in the case of single breaks, if both chromatids are broken then the break occurred in G_1 or early S phase. Two-break aberrations are structural rearrangements that include dicentrics, rings, interstitial deletions, inversions or translocations. Dicentric chromosomes, rings and acentric fragments are relatively easy to score and consequently a significant fraction of available data refers to these end points. The general mechanism that leads to two-break aberrations is the mis-joining of two single breaks. A description of two-break chromosome-type aberrations is given in Savage (1975).

A review (Lloyd and Edwards, 1983) contains a summary of dose response curves for dicentrics induced in human lymphocytes (see Fig.VI.7). The curves are well represented by linear-quadratic expressions, $\alpha D + \beta D^2$. The variation of the coefficient α for dicentrics with the LET of the radiation (α is proportional to RBE_M, the RBE at low absorbed doses) is shown in Fig.VI.8. Dose protraction (lower dose rate or fractionation) results in a decreased yield of dicentrics, in line with the assumption that single breaks (responsible for the formation of dicentrics) can be repaired; the mean half life is 1 to 2 h. For low-LET radiation (e.g. ^{60}Co) the dicentric yield per Gy is of the order of 10^{-2}. Numbers up to a hundred times larger obtain for higher LET radiations, e.g. fission neutrons (Edwards and Lloyd, 1991).

In spite of its obvious importance as a possible model for some of the stages of carcinogenesis, in vitro *cellular transformation* remains a field plagued by methodological problems and substandard reproducibility of experimental results, particularly among laboratories. At this time cellular transformation makes an

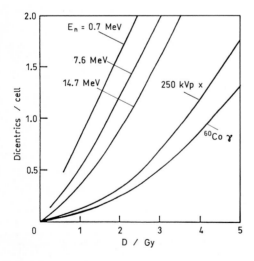

Fig.VI.7 Dose-effect curves for the yield of dicentric aberrations in human lymphocytes for different radiations (after Lloyd and Edwards, 1983).

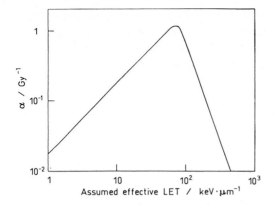

Fig.VI.8 The curves of Fig.VI.7 have been fitted to linear-quadratic expressions: $\alpha D + \beta D^2$. The dependence of α on LET is shown in this figure (after Lloyd and Edwards, 1983).

inadequate candidate for quantitative modeling other than attempting to explain some of the general, descriptive trends of the data.

Higher LET radiation is more effective in inducing cellular transformation. This is illustrated in Fig.VI.9 for C3H/10T1/2 cells exposed to x rays, and also to several mono-LET charged particles (10 to 120 keV/μm). In this case it was possible to fit

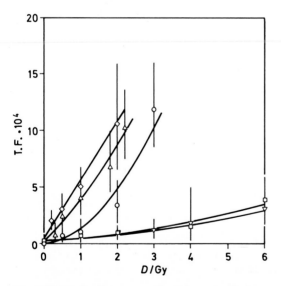

Fig.VI.9 TF, the frequency of transformed cells (per surviving cell) as a function of dose. Each curve represents a linear-quadratic fit to the data. From top to bottom the curves correspond to the following radiations: 120 and 80 keV/μm ^{3}He ions, 40 keV/μm deuterons, 10 keV/μm protons, and 250 kVp x rays (after Hei et al, 1988).

a function of the type $A[1-\exp(-\gamma-\alpha D-\beta D^2)]$ to the data; this expression reflects both the initial linear-quadratic character of the curves and the expectation of a plateau at larger doses. In these experiments a maximum RBE of about 30 (ratio of initial slopes) obtains at LET=120 keV/μm.

In the case of high LET radiation dose protraction may lead to an *increase* in the yield of oncogenic transformation. The experimental demonstrability of this effect appears to depend in a peculiar way on dose, dose rate and radiation quality; for instance, concerning this latter the "inverse dose rate effect" (as it came to be known) has been observed only in the LET range of 40 to 120 keV/μm. The microdosimetric implications of the inverse dose rate effect have been discussed by Rossi and Kellerer (1986).

Cell killing[36] is arguably one of the most often studied end point in radiation biology. It is customary to represent the results of a cellular inactivation experiment by plotting the probability of cellular *survival*, S(D), as a function of dose in a semilogarithmic representation: logS vs D. Examples of several types of survival curves are shown in Fig.VI.10.

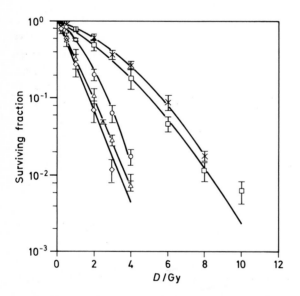

Fig.VI.10 Cellular survival data for the same radiations as those shown in Fig.VI.9; however, the order of the curves is reversed, e.g. 250 kVp x rays is the top curve (after Hei et al, 1988).

---

[36] The criterion determining "cell killing" is usually the failure of cells to divide beyond a certain limit (e.g. 32 cells).

The semi-logarithmic representation of survival curves of mammalian cells exposed in tissue culture to low-LET radiation typically have a *shoulder* (i.e. their slope steepens with increasing absorbed dose). This, as well as repair during protracted irradiation, is indication of inactivation by more than one event. With increasing LET the shoulder is reduced and it seems to be absent with such high-LET radiations as alpha particles or neutrons having energies of a few hundred keV. This is however merely due to the fact that one-event inactivation predominates in the range of the absorbed doses commonly employed . The fact that these radiations produce sublethal as well as lethal damage is made evident by the synergism observed in sequential applications of low-LET radiations (see section VI.1.5.3). The survival is then less than the product of the survival probability of the two radiations because of interaction between sublethal damage produced by both radiation.

The variation of radiation sensitivity of cells during the mitotic cycle is greater with low-LET radiation but it is also appreciable with alpha particles as demonstrated in experiments in which the survival of cells is compared when they are in different stages of the cycle (Hall et al, 1972). Hence, even if only *first order* inactivation is involved the survival curve of an asynchronous population is not strictly exponential. However, the departure from this shape may be difficult to discern. Fig.VI.11 shows the expected survival of an asynchronous population, exposed to a radiation, as calculated on the basis of measured survival of synchronized cells. The sensitivity [i.e. e(D), see Eq(VI.3)] varies by a factor of about 2 but departure from a simple exponential curve is unlikely to be detected experimentally because the precision attainable is usually not sufficient. It should also be noted that multi-event inactivation and population inhomogeneity cause opposite distortions of the survival curve.

As in the case of the other end points discussed, a number of factors influence the survival probability:

a) Radiation quality

The dependency of survival probability on radiation quality is very similar to that observed for two-break chromosomal aberrations and this has prompted attempts to connect the two end points causally. Fig.VI.12 shows the initial slope (proportional to the maximum RBE) of charged particles with differing LET values for V79 cells. The maximum inactivation effectiveness, attained at about 120 keV/μm, is followed by a decrease in RBE; this is interpreted as indication of an over-kill effect, that is more energy is deposited than necessary to inactivate the cell.

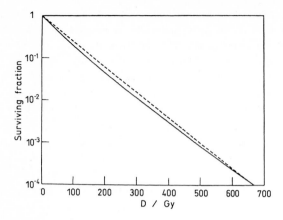

Fig.VI.11 The expected survival probability of an asynchronous population of cells exposed to a radiation as calculated on the basis of measured survival of cells synchronized in the same phase of the mitotic cycle (after Hall et al, 1972).

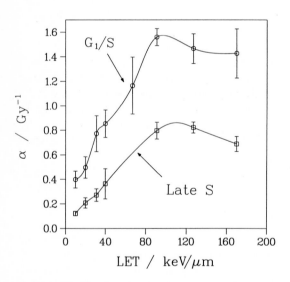

Fig.VI.12 The initial slopes, α, of survival curves measured by exposing Chinese hamster V79 cells to mono-LET charged particles (Bird et al, 1980). Cells were synchronized in late-S and G1/S phases of the mitotic cycle. The solid lines are meant to guide the eye only.

b) The temporal pattern of dose delivery

For non-exponential survival curves dose protraction results in an increase in the survival probability; in the limit of extreme protraction one obtains exponential survival curves, the slopes of which equal the initial slope of the original (non-exponential) survival curve. Exponential survival curves do not appear affected by changes in dose rate or dose fractionation. Correspondingly, high-LET radiation is less sensitive to the temporal pattern of irradiation. The process of repair (e.g. in the time interval between two separate doses) has been termed "sublethal damage repair" (SDR).

c) The phase of the cell cycle

In a random cell culture cells divide asynchronously and therefore the four conventional phases of the cell cycle (G_1, S, G_2, M) are populated at a level determined by the duration of each phase. However, when cells are synchronized in specific phases of the cycle it is noticed that cell killing depends noticeably on the phase at which radiation was received (Sinclair and Morton, 1964). The most sensitive phase is mitosis (M) and the most resistant are late S and the beginning of G_2. The ability of a cell to repair sublethal damage also depends on the phase of the cell cycle.

d) Chemical modifiers

There are innumerable chemicals that can alter the inactivation effect of radiation, whether as radiosensitizers or radioprotectors. One of the best researched sensitizers is molecular oxygen, whose presence in the cell enhances the effectiveness of radiation. For low-LET radiation this enhancement can be as high as a factor of 3 (i.e. a dose three times larger is necessary to produce the same effect in the *absence* of oxygen). However, for higher LET radiation this "oxygen enhancement ratio" (OER) becomes closer to 1. It also appears that oxygen is a *dose modifying factor*. This means that the OER is independent of dose (or, equivalently, survival level).

VI.1.4 The Theory of Dual Radiation Action

The relative biological effectiveness (RBE) of a radiation, H, relative to another radiation, L, is defined by

$$R = \frac{D_L}{D_H} \qquad\qquad (VI.26)$$

where D_L and D_H are the absorbed doses of the two radiations causing an equal effect.

The point of departure of the theory of dual radiation (TDRA) action has been the observation (Rossi, 1970) that the RBE of neutrons of moderate energies (roughly between 100 keV and a few MeV) relative to low-LET (x- or gamma) radiation depends on D, the absorbed dose of neutrons, according to a simple relation which may be written as

$$R = k \, D^{-1/2} \qquad\qquad (VI.27)$$

where R is the RBE and k is a constant. An equivalent expression is

$$R = k^2 \, D_L^{-1} \qquad\qquad (VI.28)$$

where D_L is the absorbed dose of low-LET radiation. Either of these expressions indicates that the RBE increases with decreasing level of effect, and they have been found to apply over wide ranges of doses for most, if not all, radiobiological effects that have been carefully investigated. Departures at very high or very low doses, and for intermediate-LET radiations are a necessary consequence of the theory that is based on these observations.

Investigation of the dose dependence of RBE rather than that of effect probability has two major advantages. One of these is that being based on pairs of doses required for a given effect, it is unnecessary to provide a numerical expression for the degree of effect. In many instances, such as skin damage, only an arbitrary scale can be provided and its choice determines the shape of the dose-effect curve.

The other, even more important, feature is the probable elimination of factors that modify the curve relating lesion production to absorbed dose. A dose-effect relation may depend on the number of lesions or cells involved, on the assay method chosen (e.g. the criterion for cell death) or, in general, it can be due to injury to several cells subject to various interactions. It is however not unreasonable to expect that, once the radiation-quality dependent lesions are formed, such complexities merely cause equal radiation-quality independent modifications and that the RBE merely reflects differences in the kinetics of lesion formation.

This expectation appears to be met. In the examples shown in Fig.VI.13 the simple dependence of R on D_N is not only found for such cellular effects as mutations, chromosome exchange aberrations and cell survival. It has also been observed for mammary neoplasms that are developed depending on the irradiation of several cells (Rossi and Kellerer, 1972), for life shortening (which is largely caused by a variety of tumors) and the multicellular induction of lenticular opacifications.

The TDRA has developed through 3 stages and in the following their main features and in particular their relation to microdosimetry will be briefly

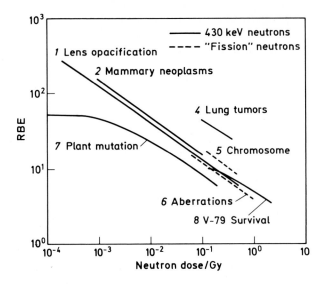

Fig.VI.13 A compilation of results showing the dependence of RBE on the neutron dose (after Kellerer and Rossi, 1972). The reference radiations are 430-keV neutrons (solid lines) and "fission neutrons" (dotted lines).

described. This needs to be preceded by definitions of the terms "lesion" and "sublesion".

A *lesion* is a cellular injury that can cause the biological effect under consideration *incoherently* (i.e. with a probability, p_s, that is independent of the presence of other lesions). Lesions can also act coherently (e.g. by combination) to form *compound lesions* that have an effect probability p_c.

Injuries that have negligible probability of causing the effect incoherently but can cause it coherently (i.e. in multiple occurrence) are termed *sublesions*.

1. The Site Model

The first formulation of the theory of dual radiation action (Kellerer and Rossi, 1972) has been termed the *site model*. It was based on what appeared to be the simplest set of assumptions. Since the yield (number) of lesions after high-LET irradiation generally increases linearly with increasing absorbed dose, and protraction (fractionation or dose rate reduction) hardly (if at all) changes the yield, it may be concluded that lesion production is at high LET due to single events and therefore proportional to the absorbed dose. Thus the average yield of lesions, ε, is given by

$$\varepsilon = \alpha_H D_H \tag{VI.29}$$

From this and Eqs(VI.27-28) it follows that

$$\varepsilon = \beta \, D_L{}^2 \tag{VI.30}$$

where $\beta = \alpha/k^2$.

The conclusion that within wide limits the dependence of cellular effect probability on absorbed dose is quadratic at low-LET and linear at high-LET suggests that the lesions responsible for radiation effects are produced by the combinations of pairs of sublesions that are produced at a rate proportional to the absorbed dose and that at high-LET pairs of combining sublesions are produced in single events while at low-LET they are produced in separate events.

In the site model it is assumed that the rate of lesion formation is proportional to the square of the number of sublesions that are produced in a site which may be considered to be the average volume in which combination occurs. Differences in the geometric distribution of sublesions within the site were ignored. It was furthermore assumed that the yield of sublesions does not depend on LET. It thus did not account for the possibility of compound lesions. The limitations of these assumptions were recognized in this simplification together with the existence of two complicating phenomena, which are transport of energy from transfer points to the matrix, and saturation. The latter effect consisting in the production of large numbers of lesions in the site should be unimportant at small absorbed doses and low LET. Although the site model must thus be regarded as only a first approximation, it can serve as introduction to the basic features of the theory.

The fundamental formula of the site model is

$$\varepsilon(z) = cz^2 \tag{VI.31}$$

where $\varepsilon(z)$ is the number of lesions in the site, z is the specific energy in the site and $\sqrt{c}$ is the average number of combining sublesions produced per unit of z. $\sqrt{c}$ thus includes not only the yield of sublesions but also the probability that a sublesion combines with another sublesion in the site to form a lesion[37].

The *average* yield of lesions $\varepsilon(z)$ is given by

$$\overline{\varepsilon(z)} = c\overline{z^2} \tag{VI.32}$$

---

[37] This relation is consistent with the finding (Powers et al., 1968) that the cross-section for cellular inactivation is proportional to the square of the LET over a wide range of the latter.

and utilizing Eq.(II.23) the dependence of ε on absorbed dose is

$$\varepsilon(D) = c(\zeta D + D^2) \tag{VI.33}$$

where ζ is the dose average of the specific energy (also written as z_D).

The first term in Eq(VI.33) accounts for lesions formed by 2 sublesions that are produced in the same event (*intra-track* action). The second term relates to the case when the combining sublesions are produced in two separate events (*inter-track* action). Apart from such complications as saturation, Eq(VI.33) applies to radiations of any LET and it determines the course of the RBE dependence on the absorbed dose of higher relative to lower LET-radiations. Utilizing Eq(VI.33) equality of effect require that

$$c(\zeta_H D_H + D_H^2) = c(\zeta_L D_L + D_L^2) \tag{VI.34}$$

where the subscripts refer to high- and low-LET radiation. With Eq(VI.26) one obtains

$$R = \frac{1}{2 D_H}\left[\sqrt{\zeta_L^2 + 4 D_H(\zeta_H + D_H)} - \zeta_L\right] \tag{VI.35}$$

When $D_H \gg \zeta_L$

$$R = \sqrt{1 + \frac{\zeta_H}{D_H}} \tag{VI.36}$$

Hence R approaches $R_\infty = 1$ at very large D_H. When $\zeta_L \ll D_H \ll \zeta_H$

$$R = \sqrt{\frac{\zeta_H}{D_H}} \tag{VI.37}$$

i.e. the RBE depends inversely on the square root of the absorbed dose of high-LET radiation.

Finally, if $D_H \ll \zeta_L$ i.e. at very low doses of high-LET radiation

$$R_M = \frac{\zeta_H}{\zeta_L} \tag{VI.38}$$

This is the maximum RBE at all sufficiently low absorbed low doses.

The condition of Eq(VI.37) cannot be met if ζ_H/ζ_L is comparable to 1 and a slope of -1/2 is then not approached. However, the condition is met in the radiobiological data shown in Fig.VI.13 and it is possible to extrapolate the (essentially) straight portions of the curves to R = 1 where according to Eq(VI.37) $\zeta_H = D_H$. Except for the effects resulting when dry corn seeds were irradiated and ζ_H is very large (~4000 Gy) values between 1.5 and 22 Gy are found. Microdosimetric measurements of the neutron beams employed in the biological experiments showed that these values correspond to site diameters between about 3 and 1 µm.

In addition to accounting for the dependence of RBE on absorbed dose, the site model was found to be in accord with various studies concerned with protraction of low-LET radiations (Rossi & Kellerer, 1973) when it is assumed that fractionation or reduction of dose rate permits repair of sublesions produced in different events.

It was, however, noted that the R_M values ranging up to 100 and more which were found in several experiments are larger than the ζ_H/ζ_L ratios for any site diameters and the possibility that c may depend on radiation quality was considered. However, the next step in the development of the TDRA was concerned with the evident possibility that the derived site diameters that are comparable to, but less than, nuclear diameters merely represent average interaction distances. It seemed likely that sublesions produced throughout the nucleus combine with a probability that is inversely related to their initial separation and that consequently equal values of ζ could result in different biological effectiveness when the relevant energy transfer point produced in an event are either more or less separated, as in a traversing track compared with a short, high-LET, track. These considerations led to the *distance model* of the theory and motivated the *molecular ion experiment* (see below).

2. The Distance Model

This formulation of the TDRA (Kellerer and Rossi, 1978) is based on fundamental concepts developed by Kellerer. It incorporates the geometrical distribution of sublesions and therefore requires not only specification of the pattern of energy transfer points as provided by the proximity function of energy transfers, (Section V.4) but also adequate information on the structure in which sublesions are formed, i.e. the matrix. The basic form of the description is similar in both cases.

Proximity functions have been discussed in Chapter V. Equivalently, the proximity function of a structure, s(x)dx, is the average volume of the structure contained in spherical shells of volume $4\pi x^2 dx$ surrounding the points in the structure (See Fig.VI.14).

In the simplest case of a uniform infinite medium s(x) is simply equal to $4\pi x^2$ but for a uniform medium with convex boundary the function is usually difficult to

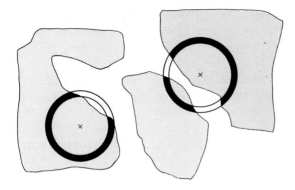

Fig.VI.14 The proximity function of a matrix, s(x)dx, is the average matrix volume contained in shells of radius x and thickness dx centered at random points in the matrix. Two such points in the matrix (light shading) are shown in the figure. The dark areas are the matrix volume that is contained in the shells.

compute. In the simplest case of a sphere one can show that

$$s(x) = 4\pi x^2 \left(1 - \frac{3x}{2d} + \frac{x^3}{2 d^3} \right) \tag{VI.39}$$

where d is the sphere diameter. With a few exceptions, s(x) for other structures has not been determined. It has been speculated, but remains unproven, that any s(x) defines a unique structure redundantly. However, even if the distribution of DNA in a cell were known a computation of only the approximate shape of s(x) would be very laborious, and the reverse procedure, i.e. a derivation of the geometry of the DNA from knowledge of s(x), appears to be a problem that is currently unsolvable. One would furthermore need to deal with an ill-defined average distribution. Nevertheless s(x) has proven to be a useful function and it can be considered to be an example of the utility of abstract concepts in practical applications.

Like t(x), s(x) is in essence a distribution of distances between pairs of points and these functions are therefore particularly suitable for the mathematical expression of the distance model. Assuming that the sublesions are produced at a rate k per unit of absorbed energy, their mean number in the cell is

$$\varepsilon_s = k\rho VD \tag{VI.40}$$

where ρ and V are the density and the volume of the matrix. The mean number of lesions is

$$\varepsilon(D) = \frac{1}{2} p\varepsilon_s = \frac{1}{2} kp\rho VD \tag{VI.41}$$

where p is the probability that a sublesion results in a lesion formation and the factor 1/2 is required because the number of combining sublesions is twice the number of the resulting lesions.

The average probability that at distances between x and x+dx from one sublesion produced at a transfer point another sublesion was produced is the product of: $s(x)/4\pi x^2$, the average *fraction* of the matrix contained in the shell; $t_D(x)$, the expectation value of the energy absorbed in that shell; k, the average yield of sublesions per unit of absorbed energy; and g(x), the combination probability. Integration over all distances yields

$$p = k \int_0^\infty \frac{s(x)\, t_D(x) g(x)}{4\pi x^2}\, dx \qquad (VI.42)$$

The proximity function at dose D may be further decomposed into two terms: consider one particular energy transfer point and a spherical shell of radius x and thickness dx centered at this point. Relevant transfer points in this shell belong either to the same event as the one to which the center relevant transfer point belongs, or to another event. Let t(x) be the proximity function for transfer points belonging to the same track. One can then write:

$$t_D(x) = t(x) + 4\pi x^2 \rho D \qquad (VI.43)$$

the second term being simply the proximity function for uncorrelated relevant energy transfer points. Hence

$$\varepsilon(D) = \frac{k^2 \rho V D}{2}\left[\int_0^\infty \frac{s(x)t(x)g(x)}{4\pi x^2}\, dx + \rho D \int_0^\infty s(x)g(x)dx\right] \qquad (VI.44)$$

Like the site model the distance model results in lesion production with linear-quadratic dependence on absorbed dose. It may be written as

$$\varepsilon(D) = K(\xi D + D^2) \qquad (VI.45)$$

with

$$K = \frac{k^2 \rho^2 V}{2} \int_0^\infty s(x)g(x)dx \qquad (VI.46)$$

and

$$\xi = \frac{\displaystyle\int_0^\infty \frac{s(x)g(x)t(x)}{4\pi\rho x^2}\, dx}{\displaystyle\int_0^\infty s(x)g(x)dx} \qquad (VI.47)$$

It is possible to express the lesion yield in terms of two functions. One of these, $t(x)$, depends only on radiation quality. The other, $\gamma(x)$, represents the probability that two energy transfer events combine to form a lesion. With

$$\gamma(x) = \frac{g(x)s(x)}{4\pi\rho x^2 \int\limits_0^\infty g(x)s(x)dx} \qquad (VI.48)$$

Eq(VI.44) may be written as

$$\varepsilon(D) = K\left[D \int\limits_0^\infty t(x)\gamma(x)dx + D \right] \qquad (VI.49)$$

If cells are irradiated with radiations of different known proximity functions, $t(x)$, and the yield $\varepsilon(D)$ is determined, $\gamma(x)$ can, in principle, be estimated on the basis of the integral equation, Eq(VI.49), when the other two functions are known. This is a difficult mathematical problem, the realization of accurately known distributions $t(x)$ is limited, and the radiobiological data have appreciable uncertainty. It has nevertheless been found possible to obtain useful information on $\gamma(x)$ on the basis of data obtained in the molecular ion experiment in which pairs (or triplets) of ions traverse cells in tissue culture with varying separation (Rossi 1979, Bird 1979, Kellerer et al. 1980, Zaider and Brenner 1984).

Fig.VI.15 illustrates the general scheme of the molecular ion experiment. In this example deuteron ions (D_2^+) are the ions that are accelerated to of few MeV. In passage through a thin foil the electron is detached and the deuterons are subject to multiple Coulomb scattering. They emerge with a known distribution of separations and pass through cells attached to the foil. The mean separation can be altered by changing the thickness of the foil. Irradiation with single deuterons of the same energy yields survival curves for "infinite" separation and, apart from minor differences in the delta-ray pattern, single ions of twice the LET of the deuterons simulate "zero" separation.

Analysis of the results yielded the functions $\gamma(x)$ shown in Fig.VI.16. Although there is some uncertainty regarding the accuracy of these curves it can be concluded that sublesion interactions occur predominantly over either very short or much longer distances; these are of the order of tens of nanometers and of micrometers. The terms *cislesions* and *translesions* have been applied to the two classes.

3. Compound Dual Radiation Action (CDRA)

The essential finding of the molecular ion experiment, which is the indication that sublesions combine over two classes of distances, has motivated attempts to

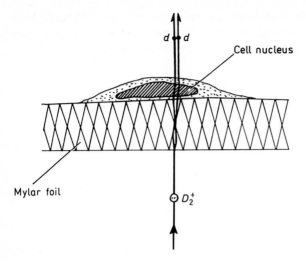

Fig.VI.15 Schematic representation of the track trajectories in the molecular ion irradiation.

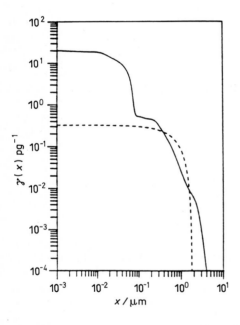

Fig.VI.16 The function $\gamma(x)$ for V79 Chinese hamster cells in late-S phase (Kellerer et al, 1980; solid line) is compared to the function $g(x)$ corresponding to a sphere of uniform sensitivity of diameter 1.8 mm (dashed line).

identify the cytological elements that are abstractly termed "sublesions" and "lesions". An evident possibility, already mentioned, are simple chromosome breaks and chromosome exchange aberrations (termed "two-hit aberrations").

Other mechanisms that may be less obvious or unknown would also be subject to the formalism of *compound dual radiation action* if they involve lesions produced by highly localized energy absorption which combine in larger regions to form compound lesions. This interpretation requires major changes in the simple assumptions of the earliest forms of the TDRA.

Chromosome breaks arise from *double-strand breaks* of DNA (*DSBs*) which in turn are due to pairs of *single-strand breaks* (*SSBs*); usually when these are produced in close proximity. Unjoined SSBs are believed to be subject to rapid repair and to be of minor importance in the biological action of ionizing radiation. DSBs are also subject to repair but this is not necessarily complete and it is obviated when a chromosome break results. Broken chromosomes may rejoin but if this does not occur they can cause deleterious effects including inability of the cell to complete more than a few mitoses. Exchange aberrations which are a misjoining of broken chromosomes can hardly be repaired. They are extremely harmful and may cause prompt death of a cell or severe abnormalities.

The formation of DSBs from SSBs would seem to be a clear example of the production of lesions by pairs of sublesions that are generated at a rate proportional to the specific energy. DSBs would then be produced at a rate that is proportional to the square of the specific energy in sites that are so small that the geometric distribution of relevant transfer points may be unimportant. It is likely that saturation is of major importance because it may require no more than two relevant transfer points to cause a DSB (see Fig.VI.5). On the other hand it is uncertain whether the number of chromosome breaks is proportional to the number of DSBs. However only single events need to be considered because multiple events are quite unlikely for any ionizing radiation when regions having dimensions of tens of nanometers are irradiated with the absorbed doses commonly applied in radiobiology (see section VI.1.2c). Hence the yield of chromosome breaks (and any other DNA damage due to highly local energy depositions) may be written as

$$\varepsilon_s = l[f(y)]D \qquad\qquad (VI.50)$$

where l[f(y)] is a *functional* of the lineal energy distribution, f(y), and represents the yield of chromosome breaks per unit of absorbed dose. According to this approach ε_s depends on the lineal energy of an event but is proportional to the absorbed dose. Hence the RBE does not depend on the absorbed dose of either radiation and is simply equal to

$$R_s = \frac{l[f(y)]_H}{l[f(y)]_L} \qquad\qquad (VI.51)$$

ε_s is a yield of *lesions* (chromosome breaks) that can cause biological effects incoherently. However, these may also combine coherently to form *compound lesions* (e.g. exchange aberrations).

The initial development of CDRA (Rossi & Zaider, 1992) employed site models. As mentioned above this may be adequate when there is interaction between a few relevant transfer points in small regions but it is an oversimplification in the case of larger regions where the distance model is more realistic. However the main conclusions of the simpler treatment should apply in either case.

Considering the yield of breaks in a larger region, D in Eq(VI.50) needs to be replaced by z, the specific energy in that region, and Eq(VI.33) which applies in the absence of CDRA, when applied to ε_c, the yield of compound lesions, becomes

$$\bar{\varepsilon}_c(D) = \frac{c}{2} l[f(y)]^2 \bar{z}^2 = \frac{c}{2} l[f(y)]^2 (\zeta D + D^2) \qquad (VI.52)$$

where $\sqrt{c}$ is now the average number of combining breaks produced per unit of z. It should be noted that y is the lineal energy in the small regions while z and ζ refer to the specific energy in the large regions.

The introduction of $l[f(y)]$ changes the RBE dependence on D_H the absorbed dose of high-LET radiation. Eq(VI.35) becomes

$$R_c = \frac{1}{2 D_H} \sqrt{\zeta_L^2 + 4 \left[\frac{l[f(y)]_H}{l[f(y)]_L} \right]^2 D_H (\zeta_H + D_H)} - \zeta_H \qquad (VI.53)$$

and the lowest value of RBE is

$$R_\infty = \frac{l[f(y)]_H}{l[f(y)]_L} \qquad (VI.54)$$

at $D_H \gg \zeta_H$, and it is equal to the constants RBE for chromosome breaks, Eq(VI.51). The maximum RBE

$$R_M = \left[\frac{l[f(y)]_H}{l[f(y)]_L} \right]^2 \frac{\zeta_H}{\zeta_L} \qquad (VI.55)$$

occurs at very low doses when $\zeta_L \gg D_H$.

The postulated mechanism of compound dual radiation action furnishes a simple explanation of differences in maximum RBE values. When R_M is relatively small (e.g. in cell killing) damage in the nanometer domain is relatively more important than at large R_M values (e.g. as often observed in cell transformation and carcinogenesis) where effects at micrometer level are more dominant.

In general a biological effect may be due to either simple chromosome breaks or exchange aberrations. If the probabilities that these lesions cause the effect are respectively p_s and p_c the effect probability becomes

$$p = 1 - e^{-(p_s \zeta + p_c \zeta_c)} \tag{VI.56}$$

provided the fraction of broken chromosomes joining in exchange aberrations is small. Depending on the relative values of p_s and p_c the RBE has intermediate values between R_s and R_c, Eqs(VI.51,53).

VI.1.5 Other Topics in Dual Radiation Action Theory

1. The equivalence of the site model and distance model

The application of *experimental microdosimetry* to radiobiology is based on the *site* model. In practice, the geometric sites that can be considered are restricted to simple shapes, such as spheres or cylinders. This is due to the fact that, as indicated in Chapter IV, proportional counters - the devices commonly used for measuring microdosimetric spectra - are actually built in such shapes. The advent of the *distance model* formulation of TDRA, prompted by radiobiological data (e.g. the molecular beam experiment, see above), has made it clear that the statistics of energy deposition in "simple" spherical or cylindrical sites can not accurately account for biological effects. The role and usefulness of proportional counter microdosimetry as a descriptor of radiation effects became thus questionable. In the following we indicate that this need not necessarily be the case; we show that: a) the site model and the distance model are only *complementary* aspects (i.e. two possible interpretations) of the TDRA formalism and that current biophysical experimental evidence can not distinguish between them, and b) for any cellular system characterized by a function $\gamma(x)$ the quantity ξ (that is the ratio α/β of the linear-quadratic dose effect relation) may be predicted from microdosimetric spectra measured in a *series of spherical sites* of different diameters. Practical applications of these ideas appear to indicate that only several sites (2 or 3) are sufficient.

Consider a cellular system that responds to a dose D of radiation according to:

$$\varepsilon(D) = \alpha D + \beta D^2 \tag{VI.57}$$

where, as usual, ε is the average yield of lesions. According to the distance model of TDRA:

$$\frac{\alpha}{\beta} = \xi = \int_0^\infty t(x)\gamma(x)dx. \qquad (VI.58)$$

On the other hand, the dose-averaged specific energy may be expressed as (see Section V.4):

$$z_D = \int_0^\infty t(x)\left[\frac{f_p(x)}{4\pi x^2 \rho}\right]dx. \qquad (VI.59)$$

Here $f_p(x)$ is the *point pair distance distribution* (ppdd) of the site where specific energy is measured (or calculated). By comparing Eqs(VI.58 and 59) it is obvious that a measurement of z_D yields the value of ξ, as in the site model, only if the sensitive site of the cell and the microdosimetric volume *have the same ppdd*. Formally:

$$\gamma(x) = \frac{f_p(x)}{4\pi x^2 \rho}. \qquad (VI.60)$$

Fig.VI.16 shows a comparison between an experimentally determined function $\gamma(x)$ and the result expected from the right-hand side of Eq(VI.60) for a spherical site of diameter 1.8 µm; the spherical site is clearly a poor representation of the geometry of the sensitive site in those cells.

Ideally, in order to preserve the role of experimental microdosimetry in radiobiology one would like to find a geometrical shape that satisfies Eq(VI.60). To the extent that such a shape exists, the site model and the distance model are equivalent. It is not obvious, however, that any function $f_p(x)$ obtained from a biological function $\gamma(x)$ with the aid of Eq(VI.60) corresponds to an actual geometric structure; nor is it evident that such a structure - if it exists - would be amenable to experimental microdosimetry (i.e. restrictions related to counter design). Instead it can be shown (Zaider and Rossi, 1988) that *any* function $\gamma(x)$ can be represented as an algebraic superposition of ppdd-s corresponding to spherical objects; thus the biophysical parameter ξ can be determined through a series of microdosimetric measurements in *spherical* proportional counters of different sizes. The mathematical formulation of this *theorem* is:

$$\xi = \int_0^\infty \left[-\frac{1}{3}\Delta^3 \frac{d}{d\Delta}\left(\frac{\gamma''}{\Delta}\right)\right] e_D(\Delta)d\Delta. \qquad (VI.61)$$

Here $e_D(\Delta)$ is the dose-averaged energy deposited in a spherical site of diameter Δ ($z_D=e_D/m$) and γ'' is the second derivative with respect to Δ.

The actual calculation of the weighting factor [in square brackets, Eq(VI.61)] is problematic because of uncertainties associated with the determination of $\gamma(x)$ in the first place. A more pragmatic approach would be to postulate from the very beginning an analytical form with only a finite number of spherical volumes, and then adjust the free parameters for the best fit to the measured ratios, α/β. For instance, Zaider and Rossi (1988) have given an example where cellular inactivation data may be reproduced with the aid of only two spherical sites of diameters 6 nm and 10 μm. Thus ξ becomes a simple linear combination of two measured quantities (e_D for each site). Although conventional microdosimetry can not be used to measure spectra in nanometer-sized sites the *variance covariance* technique (see section IV.5.2) - which measures y_D only - is ideally suited for this application.

2. The treatment of indirect effect

It is quite generally agreed that effects of radiation can be due to indirect action which is, as indicated earlier, a transport of energy to a locus in the matrix from a transfer point that is generally located outside of it. The most commonly recognized example involves the products of the radiolysis of water, in particular the hydroxyl radical, OH. This mode of energy transport will be considered here.

The time scale of radical interaction is of the order of 1 μs or less (the details of these processes are discussed in section VII.1). Nevertheless, there is a crucial difference between direct and indirect action. The former is instantaneous and occurs at the site of the resultant sublesion, while in the latter there is *spatial* and *temporal* separation between the creation of the radical and its production of a sublesion. Even in a brief interval the damage can be modified or eliminated because of radical scavenging by radioprotective agents.

In spite of the considerable complexity introduced by the diffusion of radicals it is remarkable that Eq(VI.58), implying a factorization of the distribution of absorbed energy [i.e. t(x)] and the response of the biological system [i.e. $\gamma(x)$], remains valid. However, any lesion resulting from at least one indirect hit would involve a radical which diffused some distance before producing any damage. Instead of t(x) one should then use:

$$p_d\, t_d\,(x) + p_{id}\, t_{id}\,(x) + p_i\, t_i\,(x), \qquad (VI.62)$$

where the notation refers to pairs of direct-direct, indirect-direct, and indirect-indirect hits, respectively, and the p's are constants the sum of which is equal to one. An expression for "diffused" proximity functions has been already given (section V.4.3.2); for instance, for $t_i(x)$:

$$t_i(x) = \frac{x}{2\sqrt{\pi}\,\sigma} \int_0^\infty \frac{1}{u} \left[e^{-\frac{(u-x)^2}{4\sigma^2}} - e^{-\frac{(u+x)^2}{4\sigma^2}} \right] t(u)du. \qquad (VI.63)$$

A similar relation applies to $t_{id}(x)$ but with only half the value of σ^2 because one of the members of the pair is fixed (i.e. direct hit). The critical step here is the calculation of the variance, σ^2. A simplified example is given below[38].

Let P and T represent, respectively, the protector and target molecules. If OH is the active radical an example of a molecule of the P type is dimethyl sulfoxide (DMSO). A "competition" chemical model may set up as follows (Roots and Okada, 1975):

$$OH + T \rightarrow T + H_2O$$
$$OH + P \rightarrow P + H_2O \qquad (VI.64)$$

or in terms of concentrations [...] at a time t:

$$\frac{d[T]}{dt} = k_1[T][OH]$$

$$\frac{d[P]}{dt} = k_2[P][OH] \qquad (VI.65)$$

$$\frac{d[OH]}{dt} = I - k_1[T][OH] - k_2[P][OH] - \sum_i k_i[C_i]$$

Here I is the rate at which new OH radicals are produced and the last term on the right hand side in the last equation accounts for OH radicals removed by any other species, C_i. Under steady state conditions:

$$\frac{d[OH]}{dt} = 0 \qquad (VI.66)$$

and then one can show that the mean survival time of the OH radical (at a given concentration [P]) is given by:

$$\bar{t}(P) = \frac{\eta}{k_2(1 + \eta[P])} \qquad (VI.67)$$

---

[38] As explained in section VII.1 a full stochastic treatment of the diffusion and interaction of radicals and biomolecules is generally required.

where:

$$\eta = \frac{k_2}{k_1[T] + \sum_i K_i[C_i]} \tag{VI.68}$$

The variance is then given by:

$$\sigma^2 = 2 D_{OH} \bar{t} \tag{VI.69}$$

where D_{OH} is the diffusion constant for the OH radical. Further details as well as an application of this formalism to a set of cellular survival data can be found in Zaider and Rossi (1988).

It is interesting to note that while the presence of a radioprotector would decrease the yield of sublesions, the remaining (i.e. surviving) sublesions would be in closer proximity and therefore they will produce lesions more efficiently; the reason is that when [P] is large the average diffusion time of the OH radical is shorter.

3. The effect of mixed radiation exposure

A biological system that responds to radiation according to the postulates of the TDRA must show *synergism* when exposed to more than one radiation. This means that the total number of lesions produced by the combined radiations is *always* more than the sum of the lesions produced individually by each radiation in the mixed field. Intuitively this result is evident: sublesions produced by one radiation interact not only among themselves but also with sublesions produced by all other radiations (saturation effects are ignored here). The simplest mathematical demonstration of this synergistic effect obtains by using the site model; however, the result is general.

Consider a sequential exposure to two doses, D_1 and D_2, with the subscripts referring - here and in the following - to the two radiations. The specific energies z_1 and z_2 contributed by the two radiations produce $c_1 z_1$ and $c_2 z_2$ combining sublesions, respectively. With by now familiar notations, the yield of lesions is:

$$\varepsilon(z_1, z_2) = A(c_1 z_1 + c_2 z_2)^2$$

$$\varepsilon(D_1, D_2) = \int_0^\infty \int_0^\infty \varepsilon(z_1, z_2) f_1(z_1; D_1) f_2(z_2; D_2) d z_1 d z_2 \tag{VI.70}$$

This yields (employing the more conventional α and β parameters):

$$\varepsilon(D_1, D_2) = \alpha_1 D_1 + \beta_1 D_1^2 + \alpha_2 D_2 + \beta_2 D_2^2 + 2\sqrt{\beta_1 \beta_2} D_1 D_2. \tag{VI.71}$$

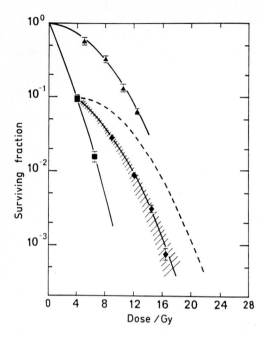

Fig.VI.17 Survival data for late-S cells irradiated with 160 keV/μm helium ions alone (■), x rays alone (▲), or graded doses of x rays after a 4-Gy dose of helium ions (◆). The curves represent fits to the data using the linear-quadratic model, $S(D)=\exp(-\alpha D-\beta D^2)$. The broken curve is identical to the x ray curve but normalized to the survival level of 4-Gy helium; this curve indicates the result expected if the two radiations act independently. The shaded area shows a 95% confidence band corresponding to the predictions of $S(D)=\exp[-\varepsilon(D)]$, with $\varepsilon(D)$ given by Eq(VI.71).

The last term in this equation represents the synergistic effect. Fig.VI.17 shows an example of cellular survival data represented with Eq(VI.71); here the approximation $S(D)=\exp[-\varepsilon(D)]$ was used. The expression, Eq(VI.71), has been generalized to include the effects of sublethal damage repair and the oxygen effect (Zaider and Rossi, 1980; Zaider, 1990; Zaider and Wuu, 1994).

4. Implications and Future Directions

The seemingly simple initial postulate of dual radiation action, which attributes the effects of ionizing radiation to lesions that are produced by pairs of damaged entities, has led to relatively complex conclusions. The loss of simplicity is accompanied with an undesirable introduction of new variables and other complications.

The function l(y), Eq(VI.50), and its dependence on physiological conditions needs to be clarified in view of apparent contradictions between available data on the dependence of DSBs and chromosome breaks on LET. It seems likely that oxygen tension and hydration are important factors. Furthermore, LET is a poor index of energy deposition especially in the small sites involved but replacement by lineal energy requires identification of the appropriate site diameter.

The subject of "repair" has also been affected because it appears necessary to consider repair of SSBs and DSBs as well as correct rejoining of broken chromosomes. In relating such phenomena to protracted irradiation it is necessary to consider the time required for the formation of exchange aberrations. A possible reason why site diameters are smaller than nuclear diameters is that the longer time interval required for combination of more distant chromosome breaks affords longer periods of repair. Thus the number of breaks produced may be larger than the number available for misjoining. Repair thus could affect the dose-effect relation even in acute irradiation.

It should again be stressed that reference to chromosome breaks and exchange aberrations may only constitute an example of other processes in which deleterious changes produced in single events may cause biological effects and combine pairwise to cause these effects as well. Other authors [e.g. Roesch (1967), Tobias (1985), Curtis (1986) and Harder (1987)] have reached similar conclusions without necessarily invoking microdosimetric considerations. However, any theories of radiobiology must be consistent with the facts of microdosimetry.

VI.1.6 DNA-lesion Theory of Radiation Action

This theory belongs to a category of models which on the basis of established dose-effect curves for a reference radiation - typically x rays or ^{60}Co γ rays - attempt to predict dose-effect curves for any other radiation, given a certain amount of physical information on the radiation field. In the case of the DNA-lesion theory this information consists of the microdosimetric spectrum, $f(z;D)$, in the gross sensitive volume (GSV) of the cell.

The building block of the theory is a quantity s representing the average number of *critical lesions* per unit molecular weight of DNA and per unit specific energy. This latter refers to the GSV of the cell. Correspondingly, the average number, $N(z)$, of critical lesions per cell resulting from an increment of specific energy z is given by:

$$N(z) = Ysz \qquad (VI.72)$$

where Y is the total mass of DNA in the cell. The biological meaning of the "critical lesions" depends on the end point considered, but it is generally associated with damage to genomic DNA. In the case of cellular survival, for instance, DNA double strand breaks are considered the critical lesions. s depends on radiation quality; this dependency is taken from experiment and parameterized in terms of the *nanodosimetric* spectrum of energy deposition in sites reproducing the gross geometry of DNA, e.g. cylinders. Günther et al (1977) suggest the expression:

$$s = \frac{1}{z_F} \sum_{j=1}^{\infty} \lambda_j f_1(j),$$ (VI.73)

where $f_1(j)$ is the *single-event* nanodosimetric spectrum, z_F is its frequency average in terms of j, the number of ionizations, and λ_j is the probability that an event consisting of j ionizations results in a critical lesion. λ_j-s are empirical parameters that need to be obtained by fitting data on DNA strand breakage.

With s thus specified, the probability of survival at dose D, S(D), is obtained by applying the site model:

$$S(D) = \int_0^{\infty} G(z) f(z; D) dz.$$ (VI.74)

G(z) represents, *for a given radiation*, the probability that an increment of specific energy, z, does not inactivate the cell. In this theory G(z) depends on radiation type because the production of critical lesions depends also on the radiation quality. Formally one writes:

$$G(z) = \sum_{v=0}^{\infty} S(v) \frac{(Ysz)^v}{v!} e^{-Ysz}.$$ (VI.75)

This expression indicates that S(v) is the survival probability of a cell with v critical lesions, and also that the number of critical lesions, v, is Poisson distributed with the mean N(z), as shown in Eq(VI.72). S(v) is further parameterized as[39]:

$$S(v) = \sum_{\mu=1}^{n} b_\mu e^{-z_\mu v}.$$ (VI.76)

---

[39] The parameterization, Eq(VI.75), is a convenient approximation to the function S(v) in that it allows closed-form integration of subsequent expressions; it is not true, however, as claimed by Günther et al (1977) that *any* function may be expressed *exactly* as a sum of exponentials unless one allows the exponents z_μ to be complex numbers, in which case one deals with the familiar Laplace transform.

It follows that:

$$S(D) = \int_0^\infty dz\, G(z) f(z;\, D) = \int_0^\infty dz \sum_{v=0}^\infty S(v) \frac{(Ysz)^v}{v!} e^{-Ysz} f(z;\, D)$$

(VI.77)

$$= \int_0^\infty dz \sum_{\mu=1}^n b_\mu\, e^{-(1-e^{-z\mu})Ysz} f(z;\, D) = \int_0^\infty dz \sum_{\mu=1}^n b_\mu\, e^{-a_\mu z} f(z;\, D).$$

The notation a_μ has an obvious meaning. Now,

$$f(z;\, D) = \sum_{m=0}^\infty \frac{(D/z_F)^m}{m!} e^{-D/z_F} f_m(z),$$

(VI.78)

and

$$\int_0^\infty e^{-a_\mu z} f_m(z) dz = \left[\int_0^\infty e^{-a_\mu z} f_1(z) dz \right]^m.$$

(VI.79)

This latter expression follows from the convolution theorem as applied to Laplace transforms. With this:

$$S(D) = \sum_{\mu=1}^n b_\mu \exp\left[-\frac{D}{z_{1F}} \int_0^\infty dz [1 - e^{-a_\mu z}] f_1(z) \right]$$

(VI.80)

Eq(VI.80) is the main result of the DNA-lesion theory of radiation action.

For practical applications of this theory one may note the following: The parameters n, z_μ and b_μ depend on the biological system only. Once determined for a reference radiation, say ^{60}Co γ rays, they may be used for any other radiation field. The quantity a_μ depends on radiation quality only through s:

$$a_\mu = (1 - e^{-z_\mu})Ys.$$

(VI.81)

Let r represent the RBE for the production of critical lesions. Then, *for any* μ, one has: $a_\mu = ra_\mu(^{60}Co)$. One concludes that given a reference radiation (and its parameters n, z_μ, b_μ) the survival probability for any other radiation obtains if the single-event microdosimetric spectrum $f_1(z)$ and r are known.

The DNA-lesion theory of radiation action discriminates between two kinds of lesions, at molecular and at cellular level; and associates with each kind nanodosimetric and microdosimetric distributions, respectively. As such, it anticipates some of the ideas put forward in the Compound Dual Radiation Action theory (see section VI.1.4). This distinction between sub-microscopic and microscopic (cellular) targets is rich in biological implications. For instance, as pointed out by Günther et al (1977), one may consider saturation phenomena *independently* at the two levels of effect postulated in the theory. Thus, the theory accommodates well the possibility that r and the RBE for cellular survival may peak at different values of LET, in line with certain experimental observations. Equally important is the practical aspect of the theory, namely, the possibility of using microdosimetry as a tool for predicting the probability of radiation effects.

VI.2 Radiotherapy

VI.2.1 General Considerations

Ionizing radiation has been applied in the treatment of diseases - and especially of cancer - for nearly one century. The radiations initially employed were relatively low energy x rays and the γ radiation from radium preparations. Subsequently, the x ray energies have been increased by some two orders of magnitude and other radionuclides (e.g. ^{60}Co) were introduced. High energy electrons, which are the charged particles produced by electromagnetic radiations, are directly applied as produced by accelerators or as beta radiation. Although at very low absorbed doses the biological effectiveness of electrons can vary by a factor of 2 or more, differences of no more than 20% are observed at the absorbed doses and at the electron energies that have been employed in therapy. Except for treatments of superficial lesions, modern therapy commonly involves electron energies of 1 MeV or more where the LET has reached the minimum value that occurs at essentially all higher energies and microdosimetric spectra and the biological effectiveness are then substantially the same. A usually minor contribution to biological effect probability can be due to nuclear interactions that result in heavy particles. Microdosimetric spectra of therapeutic beams of photons and electrons are presented in sections IV.4.3 and IV.4.4.

Radiotherapy employing heavier particles (neutrons, protons and heavier ions as well as negative pi mesons) has been motivated by two principal considerations:

• Except for protons the LET is considerably higher than that of electrons and this is advantageous because of the so-called oxygen effect: At low LET the sensitivity of aerated cells is about three times greater than that of anoxic cells. At higher LET this factor is less and approaches 1 for LET values above about 100 keV/μm. It is believed that cancer cells are generally poorly oxygenated. Normal cells are usually well aerated and at high LET the smaller oxygen effect reduces their sensitivity more than that of cancer cells.

Hence for equal damage to the normal tissues larger doses can be delivered to tumors.

- Except for neutrons, depth dose curves are such that it is possible to obtain very favorable ratios of absorbed doses in tumors relative to normal tissues. This is due to the Bragg peaks and in the case of heavier ions and, especially with pions (See Sect. IV.4.6), there is the additional advantage of a major increase of the LET in the treatment volume.

At the energies employed in therapy the LET of protons may be less than that of the electrons produced by orthovoltage x rays. Nevertheless, because of moderate production of nuclear fragments (see below) the biological effectiveness is usually near that of 250 kVp x rays. However, there are major LET-dependent RBE differences between electrons and the other ions and pions. Furthermore, while the RBE variations within irradiated volumes are usually negligible in therapy that involves electrons or protons, they are often significant with the other radiations.

Microdosimetry is the obvious method employed to document differences of radiation quality but there are often substantial difficulties in relating the physical data [usually in the form of lineal energy spectra and especially as $d(y)$] to numerical information that is useful in radiotherapy. This includes:

- The choice of absorbed doses of the high-LET radiation on the basis of the extensive experience with low-LET radiations.

- Correlation between clinical results obtained with different high-LET radiations and especially with the same radiation at different energies.

- Evaluation of RBE differences within regions exposed to an incident radiation of given type and energy.

It should be noted that the complications posed by radiation quality are greatly different in radiation protection and in radiotherapy.

In radiation protection information on the absorbed doses received and on radiation quality is usually deficient and it is not evident where dose equivalents are best determined. Furthermore, the applicable RBE (i.e. the quality factor, Q) can be considerably larger than 10. It is however independent of the low doses involved. Cellular (although not necessarily tissue or carcinogenic) effects can be assumed to be linearly related to the dose equivalent and largely independent of its temporal distribution. The accuracy requirements are quite moderate with uncertainties of 30% or less being acceptable.

In radiotherapy these conditions are reversed. Both the absorbed dose and the microdosimetric properties can be well known at locations in tumors and in other tissues. The RBE is generally less than 5. However, it depends on the absorbed dose and non-linear components of dose effect curves are important. The accuracy requirements are very high. Typical limits of 3 to 5% are necessary because of the steep rise and the proximity of the curves relating tumor control and normal tissue complications to absorbed dose (Wambersie et al., 1990).

The efficacy of therapy is commonly considered to depend on the probability of cell killing. In the approximation in which the cell survival, S, is assumed to obey the relation $\log S = -(\alpha D + \beta D^2)$, R, the RBE of higher (H) relative to lower (L) radiation is

$$R = \frac{\alpha_L}{2\beta_L D_H}\left[\sqrt{1 + \frac{4\beta_L(\alpha_H D_H + \beta_H D_H^2)}{\alpha_L^2}} - 1\right] \qquad \text{(VI.82)}$$

where the bracketed expression in the square root is $\log(S_H)$, the survival with the higher LET radiation.

According to *compound dual radiation action* and on the basis of experimental findings the β as well as the α coefficients depend on radiation quality, and, especially in view of the high accuracy demanded in radiotherapy, differences in β cannot be neglected. The RBE thus depends in a complex way on the absorbed dose, D_H, of the higher-LET radiation. Furthermore, since $\alpha_H > \alpha_L$ the recovery between, the usually applied, multiple fractions is less in the case of higher-LET radiations. Hence, the RBE in normal cells (which generally receive lower doses) is higher and recovery is less than in cancer cells. In addition scattered radiation in neutron-therapy generally has higher LET. Failure to recognize these intricacies has resulted in unacceptable damage to normal tissues in early attempts of neutron therapy.

In the absence of knowledge of the coefficients in Eq(VI.82) it is impossible to determine the absorbed doses of radiations of different LET that may be expected to cause equal effects in radiotherapy and this problem cannot be resolved solely by recourse to microdosimetry. However, microdosimetric measurements have provided useful qualitative information and they have been utilized in empirical approaches.

The microdosimetric specification of therapy beams is usually provided as d(y), the distribution of *absorbed dose* in lineal energy as measured with tissue equivalent proportional counters with simulated diameters of the order of 1 μm. The quantity y_D is an index of the biological effectiveness of the spectrum when R is proportional to y_D. However, while this applies only at absorbed doses that are so low that the quadratic term can be neglected, y_D is meaningless even then if there is appreciable saturation because in this case y_D is not proportional to RBE.

The empirical quantity y* was designed to account for saturation (Kellerer and Rossi, 1972) and is based on data of cell survival; it is given by

$$
y^* = y_0^2 \; \frac{\int\limits_0^\infty \left[1 - e^{\left(\frac{y}{y_0}\right)^2} \right] f(y)\,dy}{\int\limits_0^\infty y f(y)\,dy}
\qquad\qquad (VI.83)
$$

with y_0 typically chosen to be 125 keV/μm. In view of the high accuracy requirements y^* is hardly appropriate for heavy particle therapy. It was based on survival data obtained at much lower energies and it refers only to the α component of the survival curve. The presence of the β component necessitates conversion factors that depend on the absorbed dose, essentially as given in Eq(VI.82).

VI.2.2 Microdosimetric Distributions

There is a substantial literature dealing with microdosimetric measurements relating to radiotherapy. Many of the spectra shown in section IV.4 involve radiations employed in therapy and only a few other examples will be given here. The measurements were performed mostly with relatively small proportional counters among the types discussed in Sect. IV.3. Except for measurements of very high energy beams of protons and heavy ions, where wall-less counters disclosed contributions by individual delta rays, most of the data were obtained with walled counters.

a) Neutrons:

Fig.VI.18 shows d(y) distributions in three of the various beams employed in neutron therapy. Even with a logarithmic abscissa they differ substantially and it may be surprising that at the absorbed doses employed in radiotherapy the RBE differs by a factor that is less than 2. This may be largely ascribed to the predominant β component which varies much less with neutron energy. Beams produced in different institutions but producing neutrons in an identical reaction show minor differences in d(y) due to various construction features and target thicknesses. Substantial changes result when the collimator material is wood rather than steel (Fig.VI.19).

Small, but not insignificant, changes of d(y) with depth in irradiated phantoms have been demonstrated. Major changes occur at any depth with orthogonal distance from the beam axis. An example is given in Fig.VI.20. This illustrates the complexity in which the (lower energy) scattered neutrons produce protons of

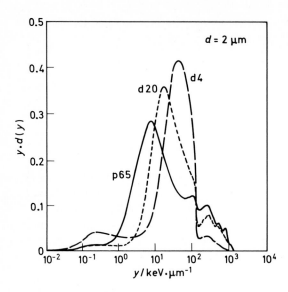

Fig.VI.18 Microdosimetric distributions for three neutron beams used in radiotherapy. The spectra indicate neutron beams produced by: 65-MeV protons incident on Be targets (solid line), 20-MeV d on Be (short-dash line), and 4-MeV d on Be (long-dash line). After: Wambersie et al (1990).

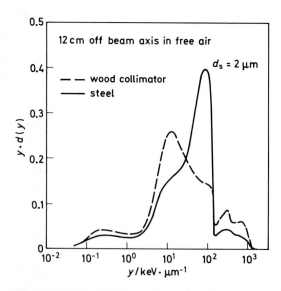

Fig.VI.19 Microdosimetric spectra for neutrons generated by 14-MeV deuterons incident on a tritium target for two types of collimator material. After Menzel (1984).

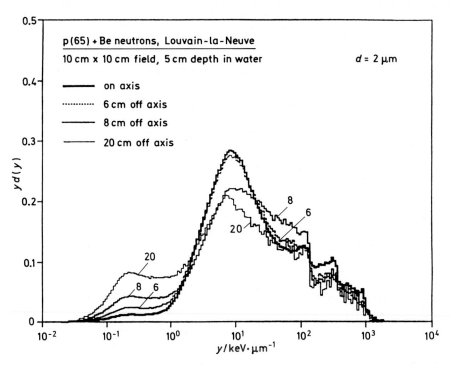

Fig.VI.20 Differences in the distributions, d(y), as a function of distance from axis of the beam in a phantom irradiated with neutrons. The neutrons are generated by bombarding a Be target with 65-MeV protons. After: Wambersie et al (1990).

higher lineal energy but produce fewer heavier recoils. In addition as the absorbed dose decreases with distance from the beam axis the relative contribution to low lineal energies by electrons becomes more important. These result from gamma radiation generated in nuclear reactions (especially capture by hydrogen). Despite these complexities neutrons are currently the high-LET radiations most widely employed in therapy. Apart from successful applications based on experience gained at particular installations it has been possible to establish intercomparisons. Based on measured d(y) spectra empirical correction functions, r(y), have been formulated (Pihet et al, 1990) with p(65)+Be as a reference with RBE=1. These distributions not only depend on the absorbed dose but also on the biological end point and they have been limited to intercomparisons between neutron therapy facilities operating at different energies. However, within these limitations this approach has been useful. The r(y) distributions calculated on the basis of measured biological responses at therapy facilities operating at various energies were employed to derive the RBE at the others with an accuracy of 3% utilizing d(y) distributions.

b) Protons

There is a degree of similarity between the therapies in which electrons are employed directly rather than indirectly as produced by photons, and the utilization of protons rather than neutrons. However in neutron therapy protons produce only the major fraction of the absorbed dose and when they are applied directly the energies required for irradiation of tumors are so high that the corresponding low LET precludes any benefit from the oxygen effect. On the other hand, like electrons, they have the advantage of delimited range. Because they scatter much less, beams of protons are far better collimated in irradiated tissues and the depth dose terminates much more abruptly.

As employed in therapy protons usually have energies in excess of 100 MeV and their LET is near that of relativistic electrons which is about 500 times less than the maximum value of 100 keV/μm at the Bragg peak. Hence the absorbed dose produced by a proton rises in a very sharp spike at the end of the range. Because of the range straggling attendant to several millions of collisions the absorbed dose in a beam which is relatively constant in a region commonly termed the "plateau" increases at what is termed the "Bragg peak" (of a beam) to less than five times the absorbed dose at the entrance. The width of the peak is of the order of centimeters. While this is optimal for such objectives as the production of sharply focussed lesions, a substantially constant absorbed dose in larger regions is commonly required in cancer therapy. The usual method for the production of a "spread Bragg peak" consists in the provision of a rapidly rotating wheel containing absorbers of a range of thickness. This technique often results in a ratio of less than two between the tumor dose and the entrance dose in a single beam. However, the steep gradients at the edges and especially at the end of the beam permit very effective treatment plans involving multiple beams.

The first microdosimetry of a proton therapy beam employed a wall-less counter (Kliauga et al., 1978) and it was possible to distinguish events due to delta rays injected by protons that pass near the counter. In Fig.VI.21a it is evident from the f(y) spectrum that the majority of the events is due to these electrons. However, the d(y) spectrum in Fig.VI.21b shows their contribution to the absorbed dose to be quite minor.

All of the spectra obtained at proton therapy installations disclose events of lineal energy in excess of 100 keV/μm which are due to nuclear reactions. This accounts for the fact that the biological effectiveness of these beams is appreciably higher than that of e.g. ^{60}Co gamma radiation.

c) Helium ions

Among the ions considered for radiotherapy applications, those of helium are in an intermediary position. Differences in the lineal energy with depth in irradiated

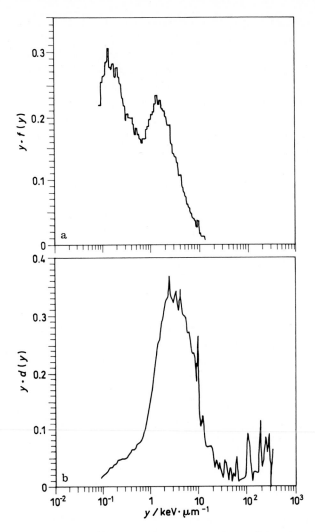

Fig.VI.21 Microdosimetric spectra for 160-MeV protons in two different representations: A) f(y) and B) d(y). Note that d-ray electrons (left peak in A) contribute little to the dose distribution in lineal energy (B).

tissues may cause some moderate change in RBE at the absorbed doses employed in radiotherapy. At the required doses there may be some slight sparing of normal tissue damage because a lower oxygen tension in cancer cells would not reduce their relative radiosensitivity as much as in photon irradiation. Adequate penetration requires considerably greater ion velocities with attendant increase of the range of delta rays that are injected into a site from external ion trajectories and this effect can be expected to be a relatively small source of error when d(y) is determined with walled counters. However the magnitude of the differences involved has not been demonstrated because the reported measurements have been performed with walled counters only.

Early measurements by Nguyen et al. (1976) were made in a beam of 650 MeV helium ions. The variable was the original "event size" Y. Multiplication by 1.5 yields values between about 9 and 22 keV/μm for y_D in a "spread Bragg peak" extending over about 5 cm before the end of the range of the ions[40]. It was recommended that prior to clinical applications the rotating step filter (that modulates the beam) be altered to attain a "biological isoeffect".

920 MeV helium ions were studied by Luxton et al. (1979). Despite the higher energy - and thus lower LET - the y_D obtained were larger ranging from about 10 to 37 keV/μm in Bragg peaks spread over 5 and 8 cm. It was suggested that the increase was due to the fact that these measurements extended to larger values of y than those reported from the earlier work. Fig.VI.22 shows the depth dose curve and d(y) at various depths for a beam when ridge filter produced a 5 cm spread.

d) Heavier Ions

A significant reduction of the differential sensitivity of oxygenated normal tissues and relatively anoxic malignant cells can be expected to occur in irradiation with ions only if they are heavier than those of hydrogen or helium. The depth of penetration needed in much of cancer therapy requires prodigious energies. Argon ions which have passed through windows, monitors and intervening air have a residual range of about 11 cm of water when they were accelerated to 18 GeV. These ions attained about 80% of the speed of light in the BEVELAC accelerator at the Lawrence Laboratory at Berkeley which could impart to any ion more than 600 MeV per amu (atomic mass unit). Very few accelerators have this capability and Berkeley has been the only installation where such ions have been utilized in experimental radiotherapy.

Various phenomena that complicate the pattern of energy deposition become important at such energies. Almost 30 percent of the ions that enter a water phantom are not argon ions but the product of interactions of the beam with intervening matter (windows, monitors, air, etc.) and this proportion increases with depth in the phantom.

Another major consideration is the prodigious range of delta rays. The theoretical maximum range of these electrons is about 8000 μm. However, the cross section for the production of very energetic electrons and the angle at which they are ejected are so small that, as calculated by Kellerer et al. (1984) most of the energy

---

[40] In the early development of microdosimetry the *event size*, Y, was defined as the energy absorbed in a sphere divided by its diameter. It is related to lineal energy, y, by: y=1.5Y.

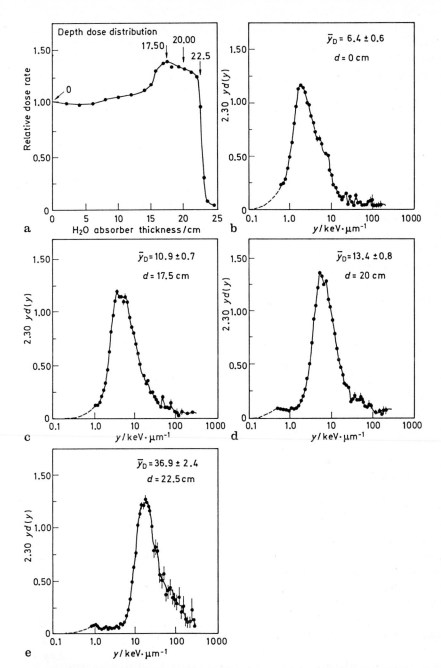

Fig.VI.22 Microdosimetric spectra for "spread Bragg peak" He ions as a function of water absorber thickness. The positions of the counter along the depth-dose curve are indicated in the figure at top left. After Luxton et al (1979).

is absorbed within a few micrometers from the ion trajectory. In microdosimetry involving a site diameter d the track core is within the site if the distance, b, between the center of the site and the trajectory is less than d/2. Measurements by Metting et al. (1988) show a sharp drop at d/2 and there is evidence of multiple delta ray traversals up to values of b that are about 3 μm when the site diameter is 1.3 μm. At larger b there are only rare events due to single electrons. These data were obtained with iron ions at 600 MeV per amu.

The range of delta rays is sufficiently large that the employment of wall-less chambers is necessary in the microdosimetry of GeV ions. Earlier measurements with nitrogen ions at lower energy (3.9 GeV at 278 MeV/amu) showed that, with site diameters of 2 μm, y_D was about 30% larger in a walled chamber as compared with a wall-less chamber (Gross et al., 1972).

Kliauga et al. (1978) employed wall-less chambers of 0.64 and 2.54 cm diameter in 10 cm enclosures to investigate the wall effect with 450 MeV/amu argon ions and 400 MeV/amu carbon ions. Technical problems (especially electrical noise) caused some uncertainties but it was evident that in the plateau y_D was substantially (~40%) larger in the bigger counter indicating that it, at least, was subject to a wall effect in an enclosure of inadequate diameter. In the smaller counter the d(y) distribution was bi-modal with the area under the lower peak about equal to that under the higher peak when a 1 μm diameter site was simulated and exposed to argon ions (and their break-up products) (Fig.VI.23a). The peaks are located at about 20 and 200 keV/μm. An appreciable fraction of the d(y) spectrum is below 10 keV/μm and due, at least largely, to delta rays. The lower peak is presumably produced by break-up products and the higher one primarily by argon ions. This distribution obtained at a depth of about 2.5 cm may be contrasted with that at the maximum of the depth dose curve (Fig.VI.23b) which has a single broad peak.

The width of the depth dose curve (full width at half maximum) is less than 1 cm and a "ridge filter" (modulating filter) was installed with the objective of creating a zone of constant biological effectiveness in a "spread Bragg peak". A single measurement in this region resulted in a distribution showing a lower secondary peak with the main peak at lower values of y (Fig.VI.23c). It seems likely that the spectrum in the "spread Bragg peak" is quite variable and that it may be exceedingly difficult to attain constant biological effectiveness in this zone. The secondary peak which is well below 100 keV/μm indicates a significant β component with attendant dose dependence of RBE.

While the microdosimetry of heavy ion beams cannot be utilized quantitatively it has been instrumental in demonstrating the extreme complexity that confronts efforts to employ them in radiotherapy.

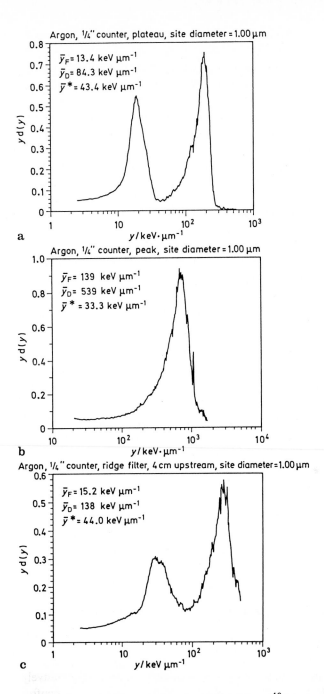

Fig.VI.23 Microdosimetric spectra for 450 MeV/amu ^{40}A ions at different positions on the Bragg curve. The two top spectra are for plateau and upstream of the Bragg peak. Second row spectra are at positions before and at the peak. The third row spectra are for the Bragg peak with a ridge filter. After: Kliauga et al (1978).

e) Negative Pions

Fig.VI.24 shows measured depth dose curves for positive and negative pions of initial momentum 78 MeV/c (Dicello et al, 1980)[41]. Also shown in this figure is the difference between the two depth-dose distributions. Because π^+ mesons do not produce stars (see section IV.4.6) this difference indicates the magnitude of the negative pion capture contribution to the dose. Microdosimetric spectra at four positions along the depth-dose curve are shown in Fig.VI.25. In the proximal (upstream) portion of the Bragg curve the microdosimetric spectra are practically identical; they reflect energy deposition by passing pions as well as equal-momentum muon and electron contaminants, unavoidably present in the beam. The main peak in the first three panels, for instance, corresponds to the stopping power of energetic (70 MeV) pions. These spectra are characteristic of low LET radiation and for these y_F is about 0.5 keV/μm. Beyond a depth of 17 g/cm^2 the shape of the spectra changes rather drastically with an increasing contribution from events of larger lineal energy, primarily protons and alpha particles generated in pion capture and by neutrons. As a relative measure of biological effectiveness for these beams one may consider either the quantity y_D or, the

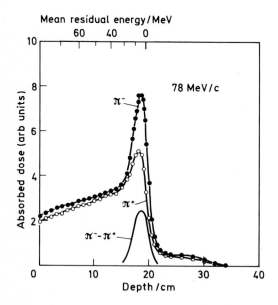

Fig.VI.24 Depth dose curves for narrow beams of π^+ and π^-, and the resulting difference. After Dicello et al (1980).

---

[41] The increase in dose in the plateau region is a result of beam optics rather than increasing stopping power for the pions.

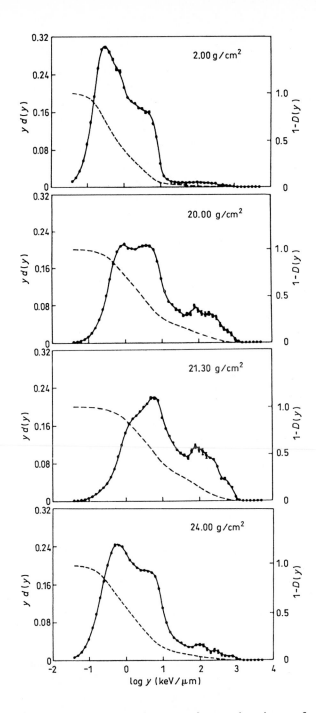

Fig.VI.25 Microdosimetric spectra for negative pions at four positions along the Bragg curve. The depth (in g/cm^2) is indicated in the panel (after Amols et al, 1975). The dashed line represents the cumulative spectrum.

saturation-corrected quantity y^*. y_D changes from 6.7 keV/μm in the plateau to a maximum of 57 keV/μm, just beyond the Bragg peak.

For patient treatment the Bragg peak of Fig.VI.24 is often too narrow and a broadening of the stopping region is needed. At Los Alamos, for instance, this was accomplished with the aid of a "range shifter". The range shifter effectively spreads out the stopping region of the negative pions. As a result, the fractional contribution of neutrons (which are not affected significantly by the range shifter) relative to high-LET radiation increases, while the contribution of these latter is diluted. An immediate consequence of this situation is the need to generate a dose distribution throughout the broad treatment volume shaped such that one obtains uniform RBE.

VI.3 Radiation Protection

VI.3.1 Quantities

a. The Quality Factor and the Dose Equivalent

The concepts of radiation protection and the corresponding recommended physical quantities have been subject to frequent modifications and further changes may well occur in the future. The earlier principle of *limitation* designed to result in negligible (or at least widely acceptable) but unspecified probability of deleterious effects, was followed by the principle of *assessment*, which is based on the assumption that the effect probabilities are proportional to absorbed dose, and on the claim that their magnitude is known (Rossi, 1984). The original primary concern with genetic effects was replaced by emphasis on radiation carcinogenesis.

The recognition that ionizing radiations differ in effectiveness and that RBE values are particularly large at the low absorbed doses of principal importance in radiation protection, necessitated formulation of a quantity that weights absorbed doses by their biological effect. After earlier formulations the International Commission on Radiation Units and Measurement (ICRU) defined the *dose equivalent*, H by

$$H = QD \qquad\qquad (VI.84)$$

as the product of the *quality factor*, Q, and the absorbed dose[42] (ICRU, 1962). Q, like RBE, is a dimensionless quantity and H therefore has the same physical dimension as D, i.e. l^2t^{-2} or in SI units, J kg^{-1}. To minimize confusion in cases where Q≠1 it was decided to admit different special names for the units of these quantities: D is expressed in gray (Gy) and H in sievert (Sv).

In accord with the usual conservative policies in radiation protection, Q is intended to approximate RBE_M (Sect. VI.1) for those effects where it is largest. Its values were related to the LET of the particles that produce D. The term "LET" refers to L_∞, or the linear stopping power with disregard of delta radiation. For many years the function Q(L) was based on a recommendation by the International Commission on Radiological Protection (ICRP, 1955) which assigned a value of 1 to particles of any LET less than 3.5 keV/μm with increases up to a maximum of 20 at 175 keV/μm and beyond. As a consequence the dose equivalent of x or γ radiation was made to be numerically equal to the absorbed dose.

A joint Task Group of the ICRP and the ICRU reexamined the subject of Q (ICRU, 1986). Its recommendations were influenced by radiobiological evidence that was obtained in the intervening period. This included RBE_M values for neutrons observed in experimental radiobiology which substantially exceeded the maximum value of Q, and RBE_M values of 2 or more for orthovoltage x rays (e.g. 250 kV), relative to hard γ radiation. It was proposed that Q be based on *lineal energy* in 1 μm sites and that Q(y) increase linearly to near a maximum of y $\approx$ 150 keV/μm with a subsequent decline due to saturation. It can be expressed by the formula (Kellerer and Hahn, 1988):

$$Q(y) = 0.3y \left[1 + \left(\frac{y}{137} \right)^5 \right]^{-0.4}$$

$$(VI.85)$$

This results in Q values that differ by a factor of about 50 between hard gamma radiation and fission neutrons.

The main advantage in the choice of y, rather than L is that it may be determined more accurately, and the selection of the site diameter of 1 μm permits relatively simple measurements. It was believed that smaller diameters could be more pertinent. However, the concept of compound dual radiation action (Sect. VI.1.3) assigns significance to local energy depositions in micrometer sites as well as in much smaller ones, and when the RBE_M is large, as it is for carcinogenesis, compound dual action dominates. In this case, the RBE is primarily determined by relative differences of energy absorbed in larger sites.

In its publication 60 (ICRP, 1991) the ICRP retained LET as the physical variable but also relegated Q(L) to lesser importance. For primary standards in radiation protection the ICRP introduced "weighting factors", w_R, that refer to the radiation incident in external exposure or emitted in internal exposure rather than to the

---

[42] Other modifying factors, often symbolized by N, were usually assigned the value of 1 and eliminated later. The original symbols for Q and H were QF and DE.

radiation at the location where D is determined. The product $w_R D$ was termed the *equivalent dose*. As stated in ICRU Report 51 (1993) the measurement of the equivalent dose is not practicable and this system will therefore not be considered here.

ICRU adopted the Q(L) relation in ICRP publication 60. (Table VI.I).

Table VI.I

Q(L) Recommended in ICRP Publication 60	
L/keV μm^{-1}	Q(L)
<10	1
10-100	0.32L-2.2
>100	300L$^{-\frac{1}{2}}$

This recommendation is somewhat similar to a formalism by Kellerer and Hahn (1988). They developed an approximation in which a relation - here denoted by Q'(L)- approaches the function Q(y) as recommended by the ICRP-ICRU task group except at low values of y or L where at 10 keV/μm Q'(L) is about 10 times larger than Q(y). As shown in Fig.VI.26, Q(L) as given in Table VI.1 differs even more and the implied equality of biological effectiveness of absorbed doses with L<10 keV/μm is in conflict with biological observations.

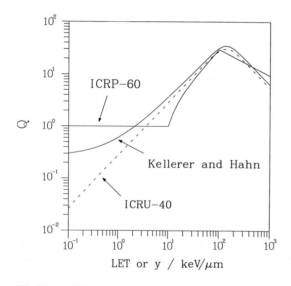

Fig.VI.26 Different formulations of the quality factor. Q(y), according to ICRU Report 40 (ICRU, 1986); Q(L) according to ICRP publication 60 (ICRP, 1991); Q'(L) approximation according to Kellerer and Hahn (1988).

In the invariable situation in which the absorbed dose is due to particles with a range of LET the dose equivalent is

$$H = \int_0^\infty Q(L)D(L)dL \qquad (VI.86)$$

The symbol D(L) refers to the distribution of absorbed dose, D, in linear energy transfer while d(L) represents a *normalized* distribution (i.e. d(L)dL is the probability that the absorbed dose is produced with L between L and L+dL). Thus D(L)=Dd(L).

Applications of these recommendations frequently employ approximations. Thus, in assessing the dose equivalent of polyenergetic neutrons, absorbed doses might be multiplied by a quality factor of 20 although this is merely the maximum value of Q which applies at neutron energies near 0.5 MeV. This procedure can lead to substantial overestimates of H and these become even greater when the same factor is applied to the total absorbed dose (e.g. as measured with a tissue equivalent ionization chamber) and nearly always includes a significant contribution by gamma radiation.

b. The Effective Dose Equivalent and Operational Quantities

In the common situation of whole body irradiation the practice of radiation protection measurements employs two principal procedures. In *area monitoring* fixed or portable instruments determine ambient radiation levels to evaluate the potential radiation exposure of persons occupying a region; and in *individual monitoring* radiation detectors are worn by persons to assess the radiation exposure actually received. The quantities to be determined are in either case dose equivalents and, like absorbed doses, they cannot be specified without reference to an actual or conceptual irradiated structure. It is a practical necessity that they relate to one or at most very few points in this structure.

In area monitoring the structure is a spherical 30 cm diameter phantom of tissue equivalent material which is termed the *ICRU sphere*. In the limitation system the maximum dose equivalent in this sphere was termed the *dose equivalent index*. For photons and neutrons of energies of usual interest this is located at depths between less than 1 millimeter and a few centimeters when these radiations are unidirectional. A detector surrounded by a tissue equivalent wall of similar thickness will then register a lower dose equivalent because of the absence of backscatter in the sphere or of secondary radiation produced in it (such as gamma radiation arising from capture of slow neutrons by hydrogen). However when the radiation is incident from a substantial range of directions, absorption by portions of the sphere can result in a dose equivalent index that is less than the dose equivalent measured by a relatively thin walled instrument. This also occurs even in a unidirectional field if, in analogy to varying orientation of exposed persons, the sphere is rotated.

The assessment system bases radiation protection on a quantity termed the effective dose equivalent, H_E, defined by

$$H_E = \sum_T w_T H_T \qquad\qquad (VI.87)$$

where the H_T are the average dose equivalents in sensitive organs in which the relative hazard (termed *risk*) of carcinogenesis or genetic effect (in case of the gonads) is w_T. This current recommendation by ICRP introduces a further complication because the effective dose equivalent received by a person is altered if he changes orientation in a constant radiation field. This complicates the directional aspects because it must be considered in addition to the angular distribution of the incoming radiation.

The *operational quantity*, H*, termed the *ambient dose equivalent* (ICRU, 1985) is based on "worst-case" assumptions: All of the generally multidirectional radiation is *aligned*; i.e. is assumed to come from the same direction, and is *expanded*, i.e. it is assumed to cover the entire ICRU sphere. The dose equivalent at a stated depth, d, in this sphere will then almost never be less (and often much larger) than the effective dose equivalent when the human body is oriented even in its most sensitive AP position (anterior-posterior irradiation) and exposed to photons or neutrons of energies up to about 20 MeV. On the basis of an extensive study involving anthropomorphic phantoms (ICRU, 1988) the recommended value of d was 10 mm. However the calculations were based on the Q(L) relation employed prior to the change recommended in ICRP publication 60. New calculations are in progress but it is unlikely that these will result in a substantial change of d.

The other quantity for area monitoring is the *directional dose equivalent*, H'(d). It was originally intended for weakly penetrating radiations (almost exclusively electrons or low energy photons) for which Q(L) is 1 and microdosimetric measurements are not required. The recommended values are d=0.07 mm for the skin and d=3 mm for the lens of the eye but larger values of d are not excluded.

In the definition of H' the radiation field is expanded but not aligned. In area monitoring it is therefore necessary in general to determine H' as a function of direction. The exposure of persons of varying orientation can be limited by H'_{max} as determined in such surveys and stationary monitors can remain oriented in the corresponding direction unless there are significant changes in the operating characteristics of the radiation source. Two advantages of this method are that in other than a unidirectional radiation field $H'(d)_{max}$ with any value of d is smaller than the (less pertinent) H*(d), and that it is the quantity that can be measured more readily.

The *personal dose equivalent* $H_p(d)$ designed for individual monitoring is defined as the dose equivalent at a depth d in the body but it may be measured by a dose equivalent meter worn on the body and covered by tissue equivalent material of thickness d.

VI.3.2 General Considerations Regarding Measurements

The early employment of microdosimetry in radiation protection was sporadic and sustained activities were not initiated until about 1980. These have been coordinated by a few groups of investigators (mostly European) who organized workshops (sect. VI.3.3 and VI.3.4).

In health physics instrumentation such advantages as simplicity and portability may often be more important than high accuracy. It should be kept in mind that all of the systems of radiation protection that have been proposed or recommended require simplifying assumptions and substantial approximations. Hence the relation between the readings of area monitors or of individual dosimeters, and the probability of detrimental effects in exposed personnel is subject to major uncertainties. Radiation protection must necessarily employ numerical values of physical quantities and exposure records of personnel are commonly given with two or even more significant figures but these are rarely realistic. The acceptable margins of error of dosimeters have been subject to various limits. The most recent recommendations are contained in an ICRU report dealing with dose equivalent measurements. An overall uncertainty of 30% is considered permissible with 10% contributed by calibration errors (ICRU, 1992). This relatively narrow range is attainable for the low LET radiations covered by that report. Considerably larger discrepancies must often be tolerated with neutrons or other radiations where mere absorbed dose measurement is insufficient and different techniques, including microdosimetry are necessary.

Determinations of the dose equivalent produced by neutrons in limited energy ranges has for many years been performed routinely with so called "adjusted" instruments. Typical examples are detectors sensitive to slow neutrons (e.g. BF_3 counters) in an enclosure of moderating material (e.g. paraffin). The accuracy of such devices which depends on the neutron energy is limited but frequently adequate in view of the various other uncertainties in radiation protection measurements. One advantage of these instruments is that they can, at least in principle, be designed so as to approximate the dose equivalents that have no physical existence, such as H*(d) in other than unidirectional irradiation.

The LET distribution of absorbed dose can be estimated from the distribution of lineal energy according to the procedure presented in Sect. IV.6. Despite the approximations involved this method is often likely to be less inaccurate in neutron measurements where "adjusted" dosimeters are nevertheless more likely to be employed in the energy range up to about 15 MeV. However at higher energies and with other, often unknown, radiations tissue equivalent proportional counters have the obvious advantage of measuring local energy deposition. At high radiation energies the finite length of tracks and their deviation from rectilinear propagation are generally unimportant. Straggling and delta ray escape increase in importance and it would be possible to apply corrections derived by Kellerer and Hahn (1988) although this refinement is usually not warranted. A study by Hartmann et al. (1981) considers eight methods of deriving H from microdosimetric data with the conclusion that any of them afford adequate accuracy for neutron energies between 0.5 and 20 MeV. The simplest of them is evaluation of $\int Q(y)D(y)dy$ with Q(y)=Q(L). However the analysis was based on the earlier form of Q(L). The relation recommended in 1991 includes the important decrease of Q when y>100 keV/μm. Derivation of the LET spectrum and application of Table VI.1 is the most direct method. However various algorithms employed with the earlier Q(L) dependence will almost certainly be adapted to the more recent form. These have included circuits of increasing sophistication that include employment of specially designed microcomputers (e.g. Kunz et al. (1990), Braby et al (1994)).

At low neutron energies the short range of recoils can cause substantial errors. The tissue region typically simulated by spherical low pressure proportional counters has a diameter of 1 μm which is the approximate range of a 100 keV proton. The standard LET analysis of the lineal energy spectrum therefore involves substantial errors at comparable and especially at lower neutron energies. Heavier recoils (e.g. oxygen nuclei) have much shorter ranges but their contribution to the absorbed dose is then small. Since the analysis in which D(L) is derived from D(y) assumes complete traversal of the counter the shorter tracks are assigned a lower LET, an error that is not on the "safe side".

Another complication arises from the fact that at low energies W (the mean energy expended in the production of an ion pair) increases and approaches infinity for proton energies below a few keV. Hence the integral $\int D(y)dy$ is less than the absorbed D when the -substantially constant- value of W pertaining to higher energy is assumed to apply. This limitation affects any measurements based on gas ionization including those made with tissue-equivalent ionization chambers. It should however be recognized that according to the concept of relevant energy transfers this is in line with a corresponding reduction in biological and other effects caused by absorbed doses of such low energy particles.

In microdosimetric measurements in radiation protection what has often been called "the dose equivalent" and will here be termed the *"counter" dose equivalent* is the dose equivalent in the sensitive counter volume which is frequently within a spherical tissue equivalent shell of 6 mm thickness. In area monitoring of neutrons of energies between 500 keV and 20 MeV this is usually a serviceable representation of the dose equivalent index or the ambient dose equivalent. It lacks the contribution from the backscatter from the ICRU sphere which causes the reading to be too low. However when the radiation is not unidirectional the relatively thin shell and the absence the self-shielding in the ICRU sphere tend to cause the reading to be too high. Efforts to extend the utilization of counters to lower neutron energies will be covered in Sect. VI.3.4.

In radiation protection microdosimetric measurements can be tedious and incapable of registering even slow changes of radiation intensity or quality because of inadequate counting rates. The measurements are often performed with 10 cm diameter counters simulating 1 μm diameter sites in tissue. In such sites Φ^*, the event frequency for fission neutrons is about 10^{-1} Gy^{-1}. Since the fluence increases as the square of the ratio of the diameters, the counter registers 10^9 events per gray. With Q=20 this represents 5×10^7 counts per sievert. Ambient dose equivalent rates in frequently occupied zones near reactors are rarely more than 10 μSv/h which corresponds to a counting rate of 500/h. Even when the spectrum is determined as a histogram with only 10 channels, a measurement lasting one hour yields a poorly defined spectrum.

Counting rates are considerably higher when Q is less. In the range 100 keV/μm >L> 10 keV/μm Q decreases linearly and, in addition, more events are required to produce a given absorbed dose. For a given dose equivalent rate the counting rate is then roughly inversely proportional to Q^2. In the extreme case of hard gamma radiation Φ^* in 1 μm sites is near 20 Gy^{-1} and, with Q=1, 20 Sv^{-1}. Hence when the dose equivalent rate is 10 μSvh^{-1} the counting rate in a 10 cm diameter counter is near 2×10^6 hr^{-1} or about 4000 times larger than at Q=20.

For a typical natural low LET radiation background of 0.1 μSv h^{-1} one thus calculates approximately 2×10^4 counts per hour. Measurements in high flying aircraft and in space vehicles often indicate counter dose equivalents that are several times larger than those at ground level. However the principal objective is an evaluation of Q which is determined by the relative contribution to the dose equivalent due to high values of y where the number of counts is usually small. In order to attain adequate counting statistics in shorter periods of time, larger counters have been employed. In measurements at SST (supersonic transport) altitudes a counter of 20 cm diameter was employed (see below).

Smaller counters may have to be employed for measurements in the environs of high energy accelerators which must contend with the almost invariable characteristic of pulsed radiation. It is frequently necessary to reduce the accelerator radiation output (which can result in changes of radiation quality) because the pulse duration is comparable to or less than the counter dead time.

This is determined by the motion of the positive ion sheath in the counter and sets a limit that is typically of the order of 1000 counts per second. It may be necessary to accept an average of less than one count per accelerator pulse and measurements must then be prolonged by a factor that is larger than the duty cycle[43] of the accelerator. If the latter is short it may then take several hours to obtain a spectrum even if the dose equivalent rate is substantial.

Most of the detectors employed in radiation protection microdosimetry have been walled proportional counters. The dose equivalent due to discrete delta rays is generally very small and can be neglected especially since the wall effect results in an overestimate of lineal energy (See Sect. IV.3.5). A variety of circuits have been employed to process the data. Because the counting rate is usually low the pulse spectrum is often analyzed in terms of a relatively small number of channels. Comparison by Kunz et al. (1990) indicated that in as few as 16 channels of varying, empirically determined width, it is possible to achieve an adequate approximation of the d(y) spectrum.

VI.3.3 Measurements of the "Counter" Dose Equivalent

A major reason for infrequency in the utilization of microdosimetry in radiation protection has been lack of familiarity and hesitance to abandon traditional methods in favor of generally more complex and often more expensive instrumentation. In exposures to the most commonly encountered low-LET radiations Q=1 and investigations of radiation quality are unnecessary. Next of importance are neutrons of energies up to a few MeV where "adjusted" instruments continue to be routinely employed. Most of the microdosimetric measurements at reactors and low-energy accelerators have been described in internal reports. However the literature includes a number of publications with some of them reported in the Proceedings of the Microdosimetry Symposia.

Microdosimetry is a major, if not the only, method to be employed in radiation protection when very high energy, or essentially unknown, radiations are involved. An early study (Rossi et al., 1962) was carried out at two installations in which protons were accelerated to energies of 0.5 GeV or more. Fig.VI.27 illustrates the obtained distributions in LET of the "counter" absorbed dose and dose equivalent (on the basis of Q(L) adopted at that time) at a particular location in the environs of one of these machines. A substantial decrease of Q with depth in a phantom was observed.

A study outside the shielding of a medical electron accelerator (Schuhmacher and Krauss, 1986) disclosed an appreciable dose equivalent due to neutrons produced in photonuclear reactions.

---

[43] The fraction of the time in which radiation is emitted.

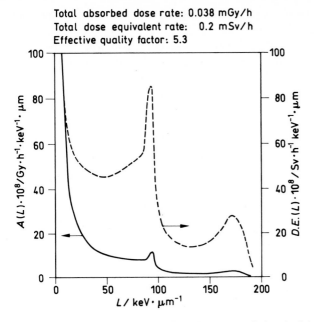

Total absorbed dose rate: 0.038 mGy/h
Total dose equivalent rate: 0.2 mSv/h
Effective quality factor: 5.3

Fig.VI.27 The distribution in LET of the "counter" absorbed dose and dose equivalent.

The elevated natural radiation background at high altitudes and in space has been subject to several microdosimetric determinations. A detailed study of the radiation at the altitudes traversed by supersonic aircraft was commissioned by the US Federal Aviation Administration and its main findings were published (FAA, 1975). In addition to an approximately 10 fold increase in absorbed dose rate, Q increased by a factor of about 2 between sea level and 20,000 m. A description of one of the instruments employed has been published (Kuehner et al., 1973).

Microdosimetric measurements have been performed in spacecraft orbiting at altitudes ranging from about 350-450 km. These include studies in the USSR Mir Station (Nguyen et al., 1990) and in US space shuttles (Badhwar et al., 1992). The instrument employed in the latter project has been described in detail (Braby et al., 1994). Substantial variations of the reported dose equivalent are due to a number of causes. The most important one is the existence of the South Atlantic Anomaly (SAA) where the belt of trapped cosmic radiation dips down to altitudes of the spacecraft. Depending on whether an orbit traverses the SAA the dose equivalent can differ by a factor of more than 2 although the traversal requires only a few minutes of the orbit period that is typically 90 min. Dose equivalent rates in excess of 1 mSv h^{-1} have been reported for the SAA with Q$\approx$2. Another source of variation are differences in the shielding at various locations in the space vehicles. Extremely large increases of radiation levels occur as a result of major solar flares but no measurements have been reported for such periods.

VI.3.4 Measurement of Operational Quantities

Efforts to determine operational rather than "counter" dose equivalents have focussed on measurements of the ambient dose equivalent. Except for the trivial case of unidirectional radiation this quantity has no physical existence because in general the radiation is not aligned. H* must be determined with an instrument having isotropic response although it refers to a very eccentric location in an irradiated phantom (the ICRU sphere).

These conditions would seem to require approximate measurements in which the response of a detector is modified by the provision of extraneous material (i.e. "adjusted" instruments) and this approach has been attempted in practice. As indicated in VI.3.1 it may be preferable to base radiation protection on the directional dose equivalent H'.

Under the aegis of the European Radiation Dosimetry Group (EURADOS) of the Commission of the European Community the utilization of tissue equivalent proportional counters (termed TEPC) was discussed in workshops (EURADOS, 1984, 1989) and intercomparisons of instruments were carried out (e.g. Alberts et al, 1988) involving exposures to neutrons of energies ranging from thermal energy to 15 MeV. The measured values were compared with values of H* obtained by calculations. The major result of these studies was that all the instruments indicated dose equivalents that were substantially less than H* at neutron energies below 1 MeV with the discrepancy reaching a factor of 4 or more. It was also noted that the absorbed dose, D, registered was less than kerma, K, although the underestimate typically only by a factor of 2.

Although the interaction processes in these instruments are complex and differ among their various configurations the results obtained can be qualitatively accounted for on the basis of considerations in Sect. VI.3.2. In the decline of D/K the increase of W at low neutron energies must be a major reason together with attenuation by the counter walls of varying thickness although thermalization of neutrons in a thick wall may have the opposite effect. This is due to the relatively low value of W when the capture reactions, $N(n,p)O$ and $H(p,\gamma)D$ produce (directly or indirectly) energetic charged particles. Larger values of D/K would therefore be expected for thermal neutrons for which this ratio was not reported.

The even lower values of H/H* were caused by incomplete traversal of the sensitive volume which was accentuated by the choice of a simulated diameter of 2 μm in most of the counters and lack of the backscattering that would occur in the ICRU sphere.

Efforts to lessen these differences have included larger wall thickness (Booz et al., 1989) and the addition of ^{3}He which, with a large cross-section produces high-LET protons and ^{3}H recoils in the (p,n) reaction (Pihet et al., 1989).

As discussed in VI.3.2 an alternative approach to area monitoring could be based on H'. Placement of proportional counters (possibly of a parallel plate type) at a depth d in a slab of tissue equivalent material would be at variance with the definition of H' in terms of the ICRU sphere. It would however be a more realistic approximation of the exposure of the human body and permit the utilization of a large sensitive region that could be occupied by several counters operating in parallel. This would increase the counting rate and thus reduce the major disadvantage of the need to determining H'_{max}.

Employment of microdosimetry to determine H_p is a desirable goal. However there are formidable engineering problems in constructing a personal dosimeter embodying a proportional counter, a pulse analysis network and recording register together with battery operated power supplies. Such a device has nevertheless been produced (Braby et al, 1994) and work on this project is continuing.

VI.3.5 Specific Quality Functions

This section details the derivation of the empirical function Q(y); for specific end points this function has been termed *specific quality factor* and denoted q(y) (Zaider and Brenner, 1985). Consider a biological system exposed to a dose D. The probability density function of specific energy z is given by (see Chapter II):

$$f(z, D) = \sum_{v=0}^{\infty} e^{-n} \frac{n^v}{v!} f_v(z) \qquad (VI.88)$$

The approach known as *hit-size effectiveness* (HSF; Bond and Varma, 1982) postulates the existence of a function, $\varepsilon(z)$, termed HSF function, which satisfies:

$$\varepsilon(D) = \int_0^{\infty} \varepsilon(z) f(z; D) dz \qquad (VI.89)$$

where $\varepsilon(D)$ is the probability of effect (a *measurable* quantity) at dose D. Both $\varepsilon(D)$ and $\varepsilon(z)$ refer to a specified biological end point. Furthermore, at low doses (the domain of interest here) one obtains from Eq(VI.89):

$$\varepsilon(D) \cong D \int_0^{\infty} \frac{\varepsilon(z)}{z_F} f_1(z) dz = D \int_0^{\infty} \frac{\varepsilon(z)}{z} \left[\frac{z f_1(z)}{z_F} \right] \qquad (VI.90)$$

In writing Eq(VI.90) we have used the property:

$$\varepsilon(0) = 0 \qquad (VI.91)$$

At low doses the sensitive site of the biological object is traversed by single charged particles or none at all. Microdosimetric events ("hits") have effectiveness $\varepsilon(z)$ to produce the observed effect, hence the name of this function. The rearrangement of Eq(VI.90) on the right hand side is instructive in relation to Eq(VI.86): the quantity in square brackets is the fraction of dose delivered by *single* events in the specific energy range [z,z+dz]. In a later publication (Zaider and Brenner, 1985) the ratio of the HSE function to specific energy has been termed *specific quality factor*:

$$q(z) = \frac{\varepsilon(z)}{z}.$$
(VI.92)

Both quantities, $\varepsilon(z)$ and $q(z)$, have the same conceptual meaning; they will be used interchangeably as deemed convenient.

The following should be noted:

a) Eq(VI.90) is a recipe for calculating a microdosimetric-based quality factor, Q, for any radiation field:

$$Q = \frac{1}{y_F} \int_0^\infty \varepsilon(y) f(y) dy,$$
(VI.93)

where lineal energy (different from z by a multiplicative constant) has been used instead. We stress again that Q in Eq(VI.93) refers to a specific end point.

b) Eq(VI.90) shows that $\varepsilon(z)$ may be obtained from the initial slope of the dose response curve. As an example, for a linear-quadratic response function:

$$\varepsilon(D) = \alpha D + \beta D^2$$
(VI.94)

one has Q=α.

The determination of $\varepsilon(y)$ may proceed as follows: In a series of experiments employing different radiations one obtains the initial slopes, α_i, corresponding to exposures to radiation "i" with microdosimetric spectrum $f(z;i)$. One then solves the integral system of equations:

$$\alpha_i = \int_0^\infty \frac{\varepsilon(y) f(y; i)}{y_F(i)} dy \quad i = 1,2,\dots$$
(VI.95)

c) Generally no analytic expression for $\varepsilon(y)$ need to be assumed. Non-parametric solutions to Eq(VI.95) may be obtained as indicated below. If the biological mechanism of radiation action is known, and one can postulate specific analytic formulae for $\varepsilon(y)$, the solution of Eq(VI.95) is clearly facilitated.

d) The assumption that a function $\varepsilon(y)$ exists at all is known as the *site model*. Specifically, it implies that there are regions in the cell (not necessarily contiguous) such that energy deposited therein uniquely determines the biological effect.

e) A critical element in determining $\varepsilon(y)$ is the actual geometric shape of the sensitive (microdosimetric) volume of the cellular system; for the notion of HSE function becomes practical only if the result of unfolding Eq(VI.95) is not excessively sensitive to the site geometry, and if simple geometries, such as spheres or cylinders, may be used as first approximations.

Several sets of data have been used to obtain functions $q(y)$ (Varma et al, 1994). For each experiment the data have been analyzed in terms of the linear quadratic model, $\alpha D + \beta D^2$, and the value of α used for the initial slope. The function $q(z)$ sought in this analysis should yield the RBE for a radiation field. The slopes may be then normalized against the slope of a standard radiation, the common choice being x rays. This may not be the best choice as results for low LET radiation appear to be plagued by large systematic errors. For instance, Lloyd and Edwards (1983) in their review of chromosome aberrations in human lymphocytes quote (their Table 2-3) measured initial slopes (α) for exposures to ^{60}Co ranging from 0.33 to 6.94, that is changing by a factor of more than 20. Further confusion arises from the lack of a standard in defining the RBE. For instance, Skarsgard et al (1967) find a maximum RBE of 4 for chromatid aberrations, by evaluating it at doses that produce 0.4 aberrations/cell. If one calculates the *low-dose* RBE (i.e. the ratio of initial slopes) the number that obtains is 235, and also the maximum in the RBE vs LET curve changes from 126 keV/μm to 189 keV/μm. To avoid these problems one may calculate the absolute value of the function $q(y)$ expressed in units of Gy^{-1} (same as α). The analysis consists of obtaining first the microdosimetric spectra for the radiations of interest and then solving the system of equations:

$$\alpha_i = \sum_{i=1}^{N} K_{ij} \, q(y_j), \quad i = 1, 2, \ldots, M \qquad (VI.96)$$

with an obvious definition for the kernel K_{ij}. Maximum entropy methodology (see Appendix V.6) has been used to solve this system of equations.

From these analyses it appears that the site sizes selected (d= 20, 100, 1000 nm) are equally useful for generating hit-size effectiveness functions. In practical terms one could then use proportional counters for measuring, say at d= 1 μm, the microdosimetric spectrum in an unknown radiation field, and use Eq(VI.90) to generate specific quality factors for the desired end point.

In terms of radiation protection, to which the hit-size effectiveness function must be ultimately applied, one would need to address questions such as: a) what is the relative contribution of a certain end point to deleterious effects in humans ?, or b) to what extent a function q(y) obtained from cellular data is applicable to (again, the more relevant) tissue or organ effects ? These two questions are clearly not independent. Concerning the former, one would like to have end-point-dependent weight factors, w_i, in order to obtain an average hit-size effectiveness function, q(y):

$$\bar{q}(y) = \sum_i w_i \, q_i \, (y) \qquad\qquad (VI.97)$$

In regard to the second question, there are indication that - at least in terms of relative effects (to which the quality factor really applies) - various levels of cellular organization may not play a very drastic role: for instance, it has been noticed that for a variety of end points (covering cellular, tissue and organ effects) RBE versus dose curves have almost identical shapes over a wide span of doses, an indication that radiation damage is recognized as such only at the very incipient stage of an otherwise long and complicated series of biological (but radiation independent) processes.

Chapter VII
Other Applications

VII.1 Microdosimetry and Radiation Chemistry

In the application of microdosimetry to radiation biology a distinction needs to be made between *direct* and *indirect* effects (see section VI.1.5.2). In the case of *direct effect* of radiation the spatial pattern of altered target molecules is expected to coincide with that of the initial energy deposits[44]. In contrast, *indirect action* depends not only on the initial microdosimetric distribution (as represented, for example, by the radical species generated by radiation in the medium surrounding the target) but also on the probability that radicals that survive interactions among themselves diffuse and collide with the target molecule. Clearly the pattern of molecules modified *indirectly* is time dependent and not identical to that of the initial track. The qualitative features of these changes are illustrated in Fig.VII.1: the four panels (Turner et al, 1983) correspond to "snapshots" of the track of a 4 keV electron taken at times ranging from 1 psec to 0.1 μsec following passage through liquid water. During this time interval more than half of the initial radical species have reacted among themselves or have been converted to unreactive molecules; and one also notes the increasingly diffuse appearance of the track - in effect a decrease in the energy density of the track. Target molecules damaged indirectly by the track will have a spatial distribution that mimics the appearance of the track at the time of the reaction. This has important biological consequences: For instance, double strand breaks (dsb) in DNA result from the pairwise combination of single strand breaks (ssb) whenever they are within approximately 10 base pairs[45]. The relative distance between ssb-s produced in indirect action depends in turn on the pattern of hydroxyl radicals, OH, assumed to be the main species attacking the DNA molecule. In a medium containing scavengers of OH radicals - as is usually the case intracellularly - only OH radicals within several nanometers of the DNA may yield ssb-s; the others would have been scavenged earlier. Because of this the time available for diffusion is

---

[44] Energy migration, for instance along a biopolymeric target, may however alter the initial energy absorption distribution.

[45] The exact number is not known.

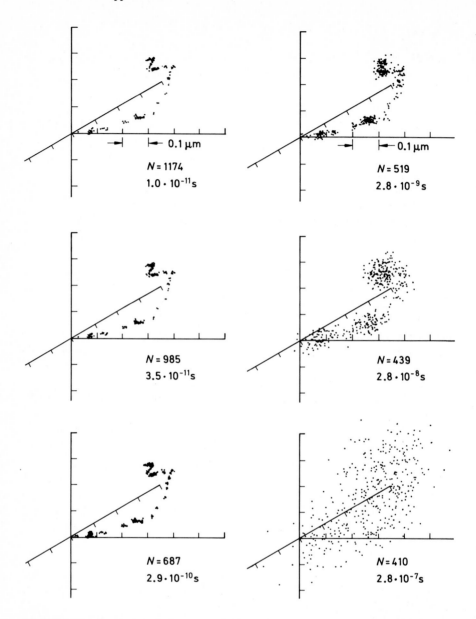

Fig.VII.1 The chemical evolution of the radicals from a 4 keV electron track as a function of time. The dots indicate radical species. Reactions among species reduce the total number of radicals. Additionally, radicals diffuse and their initial spatial disposition (top panel, left) becomes almost totally blurred after about 1 μs (bottom panel, right). The result of these two processes is a dilution in the concentration of radical species (after Turner et al, 1983).

only several nanoseconds and, paradoxically, in the presence of a scavenger the probability of induced ssb-s to yield dsb-s is larger (compare second and fourth panels in Fig.VII.1). It has been hypothesized also that the spatial proximity of damaged targets may affect their chance of being enzymatically repaired in the cell.

The study of the (usually) fast chemical processes - diffusion and interaction - following exposure of a medium to radiation is the object of *radiation chemistry*. The essential feature that differentiates radiation chemistry from "conventional" chemistry is the fact that the system being studied - ensembles of radiation induced radicals - consists of molecular species distributed *non-homogeneously* in space. Furthermore, both the yield of various species and their spatial configuration are stochastic quantities. Conceptual links between microdosimetry and radiation chemistry are thus evident, although the potential to exploit these links has been recognized only relatively recently.

The origins of modern radiation chemistry may be traced back to studies by Magee and his co-workers in the nineteen fifties (Samuel and Magee, 1956). The preferred approach was to consider a "typical" distribution of radicals, obtained as an average over possible ensembles of radicals, and then follow its evolution in time. Deterministic kinetic equations were used for this purpose. To understand the difference between this approach (still in use) and modern treatments that recognize the stochastic nature of these processes (see below) consider ensembles consisting of only one kind of species, A. Assume further that the only possible reaction is:

$$A + A \rightarrow B \qquad\qquad (VII.1)$$

At any time, t, the diffusion-controlled distribution of radicals A in space is given by (Chandrashekar, 1943):

$$\Phi(r, t)dr = \frac{1}{\left[4\pi D(t + t_0)\right]^{3/2}} e^{-\frac{r^2}{4D(t+t_0)}}. \qquad\qquad (VII.2)$$

$\Phi(r,t)dr$ is the probability that a radical initially (t=0) at the origin arrives at time t at position dr about r. In Eq(VII.2) D is the diffusion constant and t_0 is a parameter. The procedure known as *prescribed diffusion* (Kuppermann, 1961) assumes that the reaction, Eq(VII.1), does not affect the form of the distribution Φ. The probability that this reaction occurs in the time interval [t,t+dt] is then:

$$\lambda(t)dt = 2k\left[\int_0^\infty \Phi^2(r, t)dr\right] dt, \qquad\qquad (VII.3)$$

where k is the reaction constant. The average number of species A may be determined in the usual way (McQuarrie, 1967):

$$\frac{d < A(t) >}{dt} = -\frac{1}{2} \lambda(t) < A(t) > \left[< A(t) > -1 \right]. \qquad (VII.4)$$

As well as representing the name of the species, A(t) also designates its concentration at time t. The brackets <...> indicate average value. With $\lambda(t)$ given by Eq(VII.3), the differential equation, Eq(VII.4), can be solved analytically. Eqs(VII.3,4) have been generalized in many ways, for example by including more than one species or by removing the assumption of prescribed diffusion. Common to all these is, however, the tacit (and *incorrect*) assumption that the two procedures:

a) of generating an *average* ensemble, in terms of the initial distribution of species, and following its kinetics, and
b) following the kinetics of each *individual ensemble*, and then averaging the results,

are equivalent. In fact it has been shown (McQuarry, 1967) that the correct evaluation of <A(t)> is given by:

$$\frac{d < A(t) >}{dt} = -\frac{1}{2} \lambda(t) < A(t) \left[A(t) - 1 \right] >. \qquad (VII.5)$$

Eq(VII.4) can be used only if $<A(t)^2>=(<A(t)>)^2$, that is if the variance of A(t) is zero. For radiation-induced radical ensembles this is, strictly speaking, never the case.

To the extent that radicals may be assumed to interact pairwise *independently* one may relate the probability $\lambda(t)$ to the microdosimetric concept of *proximity function of distances*, f(r). Indeed, let F(r,t) represent the probability that two radicals - initially a distance r apart - interact by time t. The survival probability (i.e. the probability that no interaction took place) for an isolated pair of radicals is:

$$S(t) = 1 - \int_{0}^{\infty} f(r)F(r, t)dr, \qquad (VII.6)$$

and then, by definition:

$$\lambda(t) = \frac{1}{S(t)} \frac{dS(t)}{dt}. \qquad (VII.7)$$

This approach, originally discussed by Clifford et al (1982), has been implemented in Monte Carlo codes that deal with "realistic" simulated charged particle tracks that interact with liquid water or with biomolecules in aqueous solution (Turner et al, 1983; Zaider et al, 1994). One such code, based on the idea that pairs of radicals interact independently, is considered in the following.

The starting point of the simulation is a set containing the type and positions of all species present immediately following the passage of the ionized particle and the very fast reactions that occur in the time interval 10^{-15}-10^{-12} sec. Species are assumed to diffuse and interact pairwise; for $F(r,t)$ one may use Smoluchowski's relation (Clifford et al, 1982):

$$F(r, t) = \frac{a}{r} \, erfc\left[\frac{r - a}{\sqrt{4D't}}\right], \qquad (VII.8)$$

obtained under the assumption that two radicals interact whenever they are at a separation equal or smaller than "a" the *encounter radius*. In Eq(VII.8) erfc is the complementary error function. D' is the coefficient of relative diffusion of the two particles, that is the sum of their diffusion coefficients. The encounter radius, a, may be obtained in a first approximation if one knows the rate constant, k, between the species in the pair. This result, known as Debye's equation (Debye, 1942), is:

$$k = 4\pi a D' N_A \frac{Q}{e^Q - 1},$$
$$\qquad (VII.9)$$
$$Q = \frac{Z_1 Z_2 e^2}{a\varepsilon \, k_B \, T}.$$

Here e is the electronic charge, Z is the atomic number of each species, k_B is Boltzmann's constant, N_A is Avogadro's number, and T is the absolute temperature.

The simulation proceeds in two cyclic steps:

a) At any given time the point pair distribution of distances (proximity functions) is calculated and whenever two radicals are within encounter distance they are made to interact. Should the reaction result in a new species, it is placed in a position determined by the relative diffusion coefficients of the reactants.
b) For the remaining species one generates random encounter times (from Smoluchowski's equation) and then proceeds to eliminate sequentially pairs starting with the shortest encounter time thus generated. Pairs with one member absent, due to a previous reaction, are removed. At any time at which new species are created step a) above is performed.

A picture of the temporal evolution of the track emerges from this simulation.

An alternate approach is to simulate step-by-step the diffusion of radicals, namely having the species perform small "jumps" corresponding to a fixed time step, τ. The size of the jump, b, depends on the diffusion coefficient of the species:

$$b = \sqrt{6D\tau} \qquad\qquad (VII.10)$$

After each jump the Smolukowski's equation is used to determine the probability that reactions occurred during that time step. This approach, while more demanding in computer resources, avoids the assumption of independent pair interaction.

The general picture of water radiolysis following exposure to ionizing radiation is based on the following mechanisms involving ionization and excitation processes:

$$e^- + H_2O \rightarrow H_2O^+ + 2\,e^-$$
$$\rightarrow OH^+ + H + 2\,e^-$$
$$\rightarrow H^+ + OH + 2\,e^- \qquad\qquad (VII.11)$$
$$e^- + H_2O \rightarrow H_2O^* + e^- \rightarrow OH + H + e^-.$$

In the time interval 10^{-15}-10^{-12} sec H_2O^+ and OH^+ undergo fast reactions with water:

$$H_2O^+ + H_2O \rightarrow H_3O^+ + OH$$
$$OH^+ + H_2O \rightarrow H_3O^+ + O. \qquad\qquad (VII.12)$$

Secondary or higher-generation electrons slow down to subexcitation energies and - following thermalization - become solvated at about 0.1 psec. Thus in a medium containing only water, at about 1 psec the radiolysis species are e_{eq}, OH, H^+, H and O. This initial distribution of radicals may be characterized in terms of *proximity functions of radical pairs*. Six such examples (Zaider and Brenner, 1984) are shown in Fig.VII.2; the radicals used were induced by 22-MeV electrons for which experimental data are available. The calculated proximity functions (solid lines) are compared with Gaussian or pseudo-Gaussian distributions of the type used in prescribed diffusion applications. The parameters for the Gaussians were adjusted to reproduce experimental data (see below); however they represent merely fictitious distributions. The reactions considered in

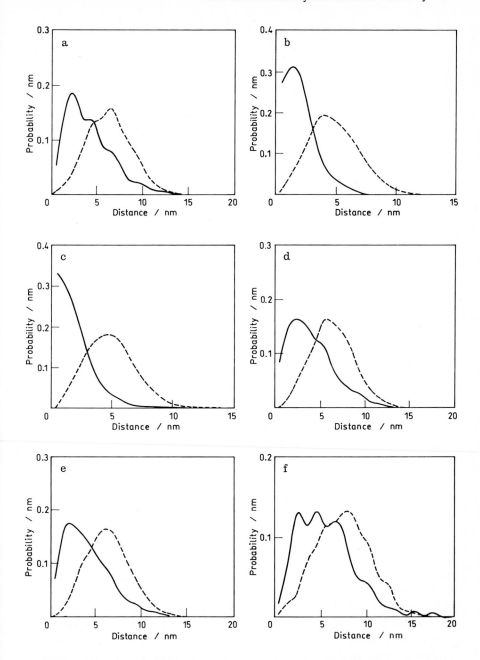

Fig.VII.2 Proximity functions for radicals induced in water by 22-MeV electrons. The panels describe the following pairs: (a) H-e_{aq}, (b) OH-OH, (c) OH-H, (d) H$^+$-e_{aq}, (e) OH-e_{aq}, (f) e_{aq}-e_{aq}. The solid lines obtain from Monte-Carlo generated particle tracks. The dashed lines represent proximity functions obtained under the assumption that radicals have a Gaussian spatial distribution; this is the usual assumption made in prescribed-diffusion calculations. After: Zaider and Brenner, 1984.

the radiolysis of pure water are shown in the Table below:

	Reaction	$k \times 10^{10}$ (dm^3/mole,sec)	a (nm)
1	$OH+OH \rightarrow H_2O_2$	0.5	0.23
2	$OH+e_{aq} \rightarrow H_2+2OH^-$	3.0	0.61
3	$e_{aq}+e_{aq} \rightarrow H_2+2OH^-$	0.5	0.40
4	$H+OH \rightarrow H_2O$	3.2	0.42
5	$H^++OH^- \rightarrow 2H_2O$	14.3	1.9
6	$H+H_2O_2 \rightarrow H_2O+OH$	0.016	0.002
7	$e_{aq}+H^+ \rightarrow H$	2.3	0.48
8	$H+H \rightarrow H_2$	1.3	0.21
9	$e_{aq}+H \rightarrow H_2+OH^-$	3.0	0.32
10	$e_{aq}+H_2O_2 \rightarrow OH^-+OH$	1.2	0.27

Figure VII.3 shows the time evolution, measured and calculated, for the average number of solvated electrons and OH expressed as G values (species per 100 eV deposited). The solid and dashed curves are, respectively, calculated results (Zaider and Brenner, 1984) and their standard deviations.

Calculations of the type described here have been performed for other radiations and/or materials, for instance a double helical DNA chain surrounded by water and exposed to low-LET radiation (Bardash and Zaider, 1994). In this case one is interested in the microdosimetry of indirect DNA damage.

As has been already mentioned, among the radiation-induced free radical species present in aqueous solution containing DNA the only species that has significant reactions with DNA is the hydroxyl radical, OH. Chemical mechanisms for specific kinds of damage have been studied quite extensively. For instance, DNA strand breakage may result from either the OH radical attacking a sugar moiety (i.e. the DNA backbone) or - as has been hypothesized - damage to a base pair may migrate to the adjacent sugar. The precise rate at which base or sugar moieties are damaged by radiation remains largely unknown and experimental evidence - when available - is only indirect. To examine these questions with the stochastic chemical approach described above one may superimpose on the radiation-induced radical tracks generated in water a three-dimensional representation of the DNA molecule consisting of nucleotides arranged, for example, in the B form of DNA. Radicals interact then not only among themselves but also with DNA moieties.

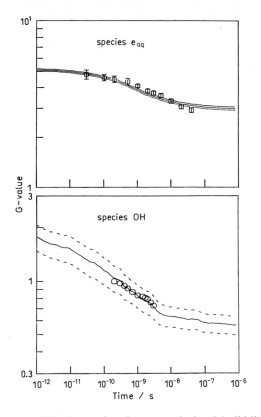

Fig.VII.3 Comparison between calculated (solid lines) and measured G values for e_{aq} and OH species induced in water by 22-MeV electrons at different times post-exposure. A 95% confidence interval for the calculation is also shown. Data from Jonah et al (1976,1977); calculations from Zaider and Brenner (1984).

As an illustration of this kind of calculation consider the interaction of 50 keV electrons with polycytidine chains[46]. The geometry of this homopolynucleotide is shown in Fig.VII.4. In this simulation electrons are directed towards the axis of a 100-nucleotide-long chain set perpendicular to their initial direction (along the z axis) and at 50 nm distance. At the doses of concern in radiobiology individual electrons may be assumed to act independently on the DNA molecule. Sugar molecules are damaged at the rate of 0.35 per incident electron track. Most of the

---

[46] A polycytidine strand consists of cytosine-sugar-phosphate units. It is usually denoted as poly-(CSP). Although a single polynucleotide chain can not have a helical configuration a system such as poly-(A-T) does have a helical structure. Furthermore, in base pairs nucleotide-nucleotide interactions (which are hydrogen bonds) may be neglected relative to ionization energies.

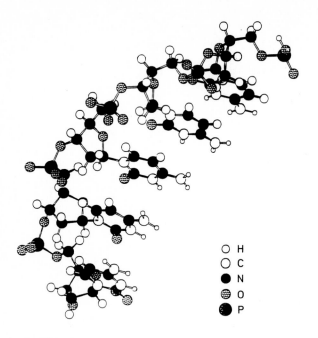

○	H
◯	C
●	N
◉	O
◐	P

Fig.VII.4 Schematic representation of the structure of polycytidine. The chain is arranged according to the B-DNA geometry.

damage (28%) consists of single altered nucleotides (see Fig.VII.5); in 5% of the tracks *two* nucleotides undergo sugar damage. When damage occurs in the sugar moiety the possibility exists that the DNA chain is interrupted. When two strands are broken in close proximity, say within 10 nucleotides, a double strand break may occur. It becomes then important to know the distribution of intervals between pairs of damaged nucleotides (again, to be considered a pair must have been produced by the *same* electron track). The results shown in Fig.VII.6 indicate that 40% of the pairs are within 10 nucleotides.

Turning now to the other DNA moiety, Fig.VII.7 shows the cumulative distributions of distances (proximity functions) between damaged *bases*. As in the case of sugar, some 40% of the distances are less than 20 nucleotides apart and therefore potential precursors of DNA double strand breaks.

An interesting question from microdosimetric viewpoint is whether one should consider the distribution of distances between *neighboring* damaged nucleotides or, alternatively, consider the distribution of distances between *any* random pair of nucleotides; this latter is, of course, the proximity function of distances. The answer may depend on the ability of the cell to repair the damage. For instance, if repair of *single* strand breaks occurs fast relative to the repair of *double* strand breaks, proximity functions would be the relevant quantity.

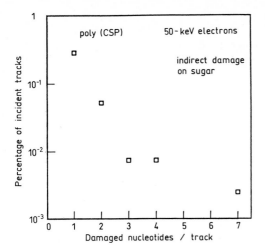

Fig.VII.5 A simulation was performed of the interaction of 50-keV electrons with poly-(CSP); the polymer was perpendicular to the electron trajectory. The plot gives the fraction of incident tracks that damage a given number of nucleotides (per track). In this figure "damage" is defined as ionization in sugar by indirect effect. After: Zaider et al (1994).

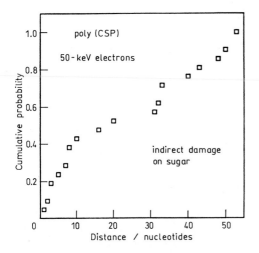

Fig.VII.6 Cumulative probability of distances between pairs of damaged nucleotides. Other details, same as in Fig.VII.5. After: Zaider et al, 1994.

One may gain further insight in indirect radiation action by restricting distributions of the type shown in Fig.VII.7 to OH radicals produced within 5 nm of the DNA chain. This may be typical of damage in *cellular* DNA where distant OH radicals are scavenged before attacking the DNA. As expected (see Fig.VII.8)

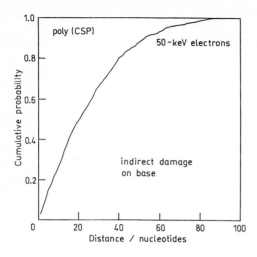

Fig.VII.7 Same as in Fig.VII.6 but for indirect damage on base.

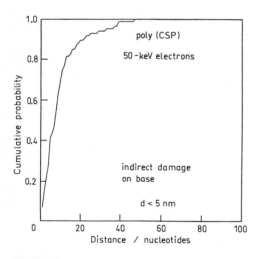

Fig.VII.8 Same as in Fig.VII.7; however, to simulate a cellular environment (where OH radicals are scavenged by various intra-cellular debris) indirect attack on DNA was restricted to OH radicals that are produced within 5 nm of the central axis of the DNA chain.

the proximity functions cover a significantly smaller range of distances: 90% of the distances are now within 20 nucleotides. The *absolute* number of pairs is, however, smaller and therefore the yield of DNA double strand breaks need not increase.

VII.2 Radiation Effects on Microelectronics[47]

Typical modern microelectronics circuits contain logical or memory elements that consist of p-n junctions. When these circuits are exposed to ionizing radiation, the resulting energy deposition (in the form of charged ion pairs) may "upset" the state of a logical unit by switching it to the opposite binary state. This is termed a *single event upset*   (SEU). SEU belong to a more general class of radiation induced phenomena termed *single-event effects* (SEE) that contain also the so-called hard core errors, that is errors that are not reversible. The application of *microdosimetry* to radiation effects on microelectronics is a direct result of the current trend in the electronics industry towards manufacturing increasingly smaller devices; and the fact that, as a result, the amount of charge necessary for inducing a SEE becomes correspondingly smaller. For instance, it has been estimated (Peterson and Marshall, as quoted in McNulty, 1990) that from the 1970's the density of junctions per unit area of chip has increased by two order of magnitude for each 10-year period while the switching energy has *decreased* at the same rate. Currently, the switching energy is of the order of 10^{-15}J, which is the equivalent of less than ten keV. This development had two main consequences: a) the fluctuations in energy deposition (and therefore in the charge induced) have become the major factor that determines whether an SEE occurs, and b) the probability that an electronics component is traversed by a charged particle track has become smaller (roughly, inversely proportional to the square of the dimension of the device). Information on these two points can be obtained with microdosimetric techniques.

Typical linear dimensions of junctions range from 1 μm to 100 μm. At the radiation fluxes encountered in the space environment[48], the probability that a device is traversed by more than one particle is negligible. Fig.VII.9 shows the spectrum for the cosmic ray fluence of charged particles at solar minimum (Simpson, 1983). It implies that in deep space the probability *per year* that a device with cross sectional area of 100 μm^2 will be traversed, for instance, by a 10 MeV/nucleon α particle is about 1 in 1000. Microdosimetric evaluations need to consider therefore single events only.

Two types of soft errors are recognized: bit upsets in memory cells and transient signals in the logic circuits. Both of these are thought to arise when a certain

---

[47] This section is based on material kindly provided to us by Dr. P. McNulty.

[48] The main concern for radiation effects on microelectronics is for circuitry flown on satellites or on missions in deep space. However, logic errors have been detected at sea level as well; they have been attributed to cosmic-ray-induced muons, pions and neutrons, and also to α particles emitted by radioactive isotopes present in the materials used in the fabrication of electronics components. This latter source of errors has been substantially reduced through repackaging and redesigning (Dicello, 1987).

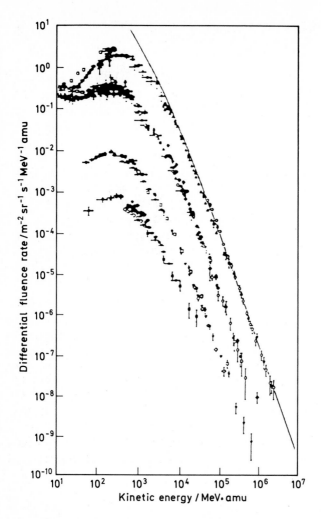

Fig.VII.9 The distribution in energy of galactic protons, He ions and C ions (from top to bottom) at solar minimum. After: Simpson (1983).

amount of charge, termed *critical charge* Q_c, is deposited in the junction. In microdosimetric language this corresponds to the *site model* assumption, specifically:

$$E(D) = \int_0^\infty e(z) f(z; D) dz \qquad (VII.13)$$

where E(D) is the yield of single-event errors in the junction after exposure to dose D, f(z;D)dz is the microdosimetric distribution in specific energy at dose D, and e(z) is the probability of effect at a given z. The idea of a critical charge means that e(z) is a step function, i.e. a threshold model.

Q_c depends on the size and geometric shape of the device. It has been estimated empirically that for a variety of devices one may represent the relationship between Q_c and d (a typical dimension for the device) as follows:

$$Q_c = 0.023 \, d^2 \qquad\qquad (VII.14)$$

The numerical constant reflects the units chosen: μm for d and pC for charge. It is then a matter of knowing the geometry of the sensitive volume of the device, and the microdosimetric spectrum of the radiation of interest, to obtain the yield of SEU per unit dose. Calculations reported to date on radiation effects on microelectronics all make use of the *track segment* assumption. As already stated, this implies that the size of the sensitive volume is much larger than the radial extension of energy deposition by single particle tracks. As junctions become smaller, the track segment assumption becomes more difficult to justify. A simple, "track segment"-type of calculation is given in the Appendix to this section.

More realistic calculations need to take into account the actual geometry of the charge-collecting volume of the exposed device, and also the fact that not all charged particles fully traverse the junction (much less at a constant LET). The simplest device one may consider in this respect is an n-p junction; schematically it consists of two silicon layers doped in a such a way as to create, at their interface, a depletion layer consisting of regions of positive and negative charge. As a charged particle traverses the junction, induced charges (electron-hole pairs) move along the electric field lines towards opposite ends of the depletion region. This results in a change of voltage across the junction and, if part of a microelectronic circuit, a potential SEU. If the junction is fully depleted, for instance by applying a reverse bias on the diode, the sensitive region may be approximated with a flat disc or parallelepiped[49]; this is because the electric field is confined to the depletion region. Most microelectronics junctions are only partially depleted and the shape of the volume from which charges are collected, following for instance the passage of a charged particle, is more complicated. A schematic illustration of this volume is shown in Fig.VII.10. One notes the depletion region (between the p- and n-doped layers), a funneling region which is a particle-induced extension of the depletion region, and a region from which charges may be collected by diffusion. It is possible that the size and geometry of the funneling and diffusion regions depend on the LET of the particle or, more precisely, the amount of energy deposited locally. If confirmed experimentally, this may invalidate the use of the site model of microdosimetry because the shape of the sensitive volume will somehow depend on specific energy. In theoretical

---

[49] Solid state particle detectors work on the same principle.

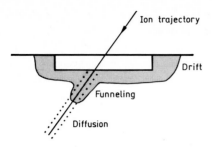

Fig.VII.10 Schematic representation of the traversal of a junction by a charged particle (see text). After: McNulty (1990).

estimations of the SEU rate it is customary to replace the sensitive volume with a regularly-shaped (e.g. parallelepiped), "effective" volume under the assumption that charge deposited in the two structures will be largely equivalent.

The shape of the microdosimetric spectrum used in the simplified calculation in the Appendix does not depend on the dimensions of the junction. This is a result of the spherical shape chosen, and also because the track segments representing the particles have been assumed to traverse fully the junction without changing their LET. If particles are allowed to originate and/or terminate inside the junction, as would be the case for neutron or proton induced spallation products, and if furthermore the continuous slowing down approximation is applied, the shape of the sensitive volume in the junction may affect the resulting microdosimetric spectrum and, consequently, the SEE yield. Fig.VII.11 shows $d(y)$ spectra for 1-MeV neutron spectra in two square parallelepipeds of 0.2 μm

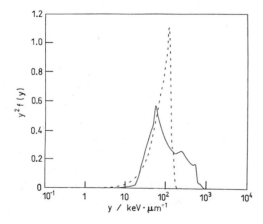

Fig.VII.11 Lineal energy spectra for a rectangular parallelepiped of 2 μm^3 volume (solid line) or 20 μm^3 volume (dashed line) exposed to 1-MeV neutrons (after Zaider et al, 1989).

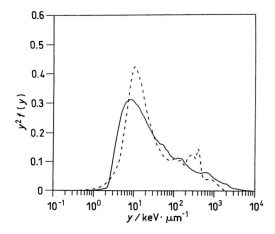

Fig.VII.12 Comparisons between d(y) spectra in parallelepipeds of volume 2 (solid line) and 20 μm³ (dashed line). Neutron energy is 14 MeV (Ibid).

(solid line) and 2 μm (dashed line) height and 10 μm² cross-sectional area; events in the range of y-values between 200 and 1000 keV/μm are all but eliminated in the larger volume. However, a similar calculation for 14-MeV neutrons shows a much less pronounced effect (Fig.VII.12). A similar trend obtains if the *volume* of the junction is kept fixed but the shape is changed. Fig.VII.13 shows two spectra for 1-MeV neutrons incident on a sphere (solid line) or spheroid of eccentricity 10 (dashed line), both of volume 2 μm³. The same calculations for 14-MeV neutrons are shown in Fig.VII.14. The conclusion from these figures is that only at lower neutron energies can the shape of the detector be used to significantly control the fractional contribution of high-LET events (i.e. those that deposit the critical

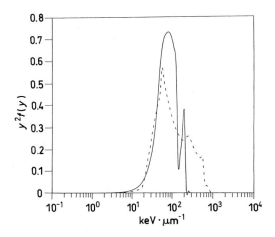

Fig.VII.13 Comparison between d(y) spectra in a sphere (solid line) and a parallelepiped (dashed line) of equal volume (2 μm³). Neutron energy is 1 MeV (Ibid).

amount of charge, Q_c); at higher energies the actual geometry of the exposed device will have only a modest effect on the overall shape of the spectrum.

SEU may be generated by both charged and uncharged particles (neutrons, γ rays). In deep space SEU production results mainly from the traversal of the junction by the heavy ions that make up the *galactic cosmic ray (GCR)* spectrum. In low-earth orbits proton-induced spallation reactions dominate the SEU spectrum. Spallation results in a break-up of the target nucleus (e.g. silicon) and the ejection of low energy, heavily ionizing charged fragments. These latter will clearly not be describable in terms of the simple model outlined above and one needs to include more sophisticated distributions of chord lengths, for example *internal randomness*. To calculate charge deposition from spallation products one needs to perform particle transport calculations not only in the sensitive volume but also in a surrounding region, the dimensions of which are equal to the longest range of the secondary spallation products. A similar situation occurs for SEU generated by neutrons; a detailed description of this case has been given by Bradford (1980) using the mathematics of geometric probability.

There have been numerous real-life examples of SEU affecting space missions. McNulty (1990) quotes the case of the Galileo spacecraft where the discovery of radiation-sensitive components has resulted   in the postponement of the launch and $15M worth of changes. Similarly, the Hubble Space Telescope appears to be affected, in addition to its publicized problems, by SEU-s at a rate much higher than anticipated.

As the need for increased speed and reduced power consumption in electronics associated with space exploration will grow (with the immediate result of further

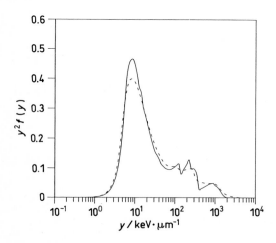

Fig.VII.14 Same as in Fig.VII.11 but for a spheroid of eccentricity 1 (solid line) or 10 (dashed line). In both cases the volume is 2 μm^3. Neutron energy is 14 MeV (Ibid).

miniaturization of electronics components), some of the assumptions currently made in calculating the rate of SEU induction will need to be relaxed; and the role of microdosimetry in radiation effects on microelectronics will correspondingly grow. Prime candidates for more refined evaluations are: a) abandoning the step-function representation of the charge necessary for producing an SEU, b) removing the track segment approximation, and c) considering the effects of specific-energy-dependent microdosimetric volumes.

VII.2.1 Appendix: Example

Consider charged particles of constant LET (=L) traversing a spherical volume of diameter d along random chords. According to Eq(V.67) the microdosimetric distribution of specific energy is:

$$f_1(z) = \frac{2z}{z_0^2}, \quad z \in [0, z_0] \qquad \text{(VII.15)}$$

where

$$z_0 = \frac{8L}{3\pi \, \rho (\bar{l})^2} \qquad \text{(VII.16)}$$

We denote by $\bar{l}$ the mean chord length in the sphere (= 2/3 of the diameter), and set $\bar{l}$ equal to d of Eq(VII.14). ρ is the material density; for Si ρ=2.32 g/cm^3. Let z_c denote the critical specific energy for a given junction; the probability to deposit specific energy in excess of z_c per particle traversal (microdosimetric event) is:

$$p = 1 - \left(\frac{z_c}{z_0}\right)^2. \qquad \text{(VII.17)}$$

Using the fact that, on average, 3.6 eV are required in Si to produce an ion pair[50], one obtains with the aid of Eq(VII.14) the numerical result:

$$p = 1 - \left(0.345 \, \frac{\bar{l}}{L}\right)^2 \qquad \text{(VII.18)}$$

To express this result in terms of SEU probability *per unit dose* we multiply p by $1/z_F$, the probability of particle traversal per unit of absorbed dose. Working out

---

[50] For GaAs the number is 4.8 eV per ion pair.

the numerical constants one obtains:

$$\alpha(Gy^{-1}) \;=\; \frac{p}{D} \;=\; 2x\,10^{-24}\; \frac{(\bar{l})^2}{L}\left[1 - \left(0.345\,\frac{\bar{l}}{L}\right)^2\right] \qquad (VII.19)$$

$\bar{l}$ and L are in μm and keV/μm, respectively. For this particular geometry (a sphere) particles for which L/l<0.345 do not induce errors; for instance, in a 10-μm junction traversed by fast electrons (L= 0.3 keV/μm) or energetic protons (L ≈ 1 keV/μm) no soft errors are induced[51]. The quantity α(L) may be considered a *quality factor* for SEU induction. For a radiation field with a spectrum of LET values one would use:

$$\bar{\alpha} \;=\; \int_0^{\infty} \alpha(L)\Delta(L)dL \qquad (VII.20)$$

where Δ(L)dL is the dose probability distribution in LET.

VII.3 Microdosimetry and Thermoluminescence

The response of thermoluminescent dosimeters (TLD) to ionizing radiation offers another example of a non-biological system that can be understood in terms of microdosimetric concepts of energy deposition. The general properties of TLD-s and the mechanisms by which they respond to radiation have been reviewed by Horowitz (1981). Briefly, the TLD material (e.g. LiF, $Li_2B_4O_7$, NaCl) contains lattice defects[52] that provide trapping sites for the electrons raised by radiation from the valence to the conduction band, as well as for the holes left behind. The energy levels corresponding to the trapping sites are in the forbidden gap and are therefore metastable. Trapped electron sites are called F-centers and trapped hole sites are called V-centers. The passage of a charged particle track through the TLD results in pairs of F and V centers along its trajectory. When the TLD is heated some of the trapped charge carriers are freed and combine pairwise (electron-hole) to produce a de-excitation photon. The optical detection of these photons is taken as a measure of the number electron-hole pairs generated by radiation in the TLD. Fig.VII.15 shows data by Li et al (1984) for the TLD

---

[51] Protons induce soft errors indirectly through the spallation products that result from nuclear reactions.

[52] Example of such defects are: impurity atoms, lattice vacancies, interstitial atoms or vacancy-impurities complexes.

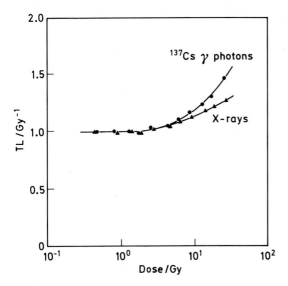

Fig.VII.15 Thermoluminescence per unit dose for LiF exposed to ^{137}Cs gamma rays and 300 kVp x rays. After: Li et al (1984).

response *per unit dose, ε(D)/D,* as a function of dose, D; in these experiments LiF was exposed to ^{137}Cs photons and 300-kVp x rays. The initial (flat) portion of the response curve indicates proportionality between the TLD response and dose, and in this range of doses TLD is usable as a conventional dosimeter. The general shape of the dose-response curves corresponds to a linear-quadratic function of dose:

$$\varepsilon(D) = \alpha D + \beta D^2. \qquad (VII.21)$$

where α and β are constants and $\varepsilon(D)$ is the TLD response. However, the theoretical discussion of these data needs to be altered because it was not recognized that in this case the two "sublesions" are produced in pairs (electrons and holes) at each relevant transfer point (RTP). Consequently, if only pairs originating at the same RTP combine the yield of photons should be the same for x and gamma radiation (except for minor differences in W).

At very high doses (of the order of 100 Gy) electrons and holes produced at different RTPs might combine. In this case the more diffuse (random) RTPs of gamma radiation would be more effective than the more non-uniform pattern of x rays. This may be similar to the "paradoxical RBE" dependence (Zaider and Rossi, 1992) observed in radiobiology at very high absorbed doses.

Further detailed studies are required to attain a quantitative accounting for the data shown in Fig.VII.15.

References

Adams, J.H. (1983) The variability of single event upset rates in natural environment. IEEE Trans. Nucl. Sci. NS-30: 4475-4480.

Alberts, W.G., E. Dietz, S. Guldbakke, H. Kluge, and H. Schuhmacher (1988) Radiation Protection Instruments Based on Tissue Equivalent Proportional Counters: Part II of an International Intercomparison. Physikalische Technische Bundesanstalt PTB-FMRB-17 Bericht 177.

Amols, H.I., J.F. Dicello, and T.F. Lane (1976) Microdosimetry of negative pions. In J. Booz, H.G. Ebert, and B.G.R. Smith (eds): Proceedings of the Fifth Symposium on Microdosimetry. Luxembourg: Commission of European Communities, p. 911.

Amols, H.I., and P.J. Kliauga (1985) Microdosimetry of 10-18 MeV electrons and photons using walled and wall-less detectors. Radiat. Prot. Dos. 13: 365-368.

Armstrong, T.W., and K.C. Chandler (1973) SPAR, a FORTRAN program for computing stopping powers and ranges for muons, charged pions, protons, and heavy ions. In : ORNL report. Oak Ridge, Tenn.: ORNL.

Badhwar, G.D., L.A. Braby, F.A. Cucinotta, and W. Atwell (1992) Dose Equivalent Rate and Quality Factor in SLS-1. Nucl. Rad. Meas. 20/3: 447-451.

Bardash, M., and M. Zaider (1994) A stochastic treatment of radiation damage to DNA from indirect effect. Radiat. Prot. Dos. 52: 171-176.

Barkas, W.H. (1963) Nuclear Research Emulsions. New York: Academic Press, p. 372.

Baverstock, K.F., and S. Will (1989) Evidence for the dominance of direct excitation of DNA in the formation of strand breaks in cells following radiation. Int. J. Radiat. Biol. 55: 563-568.

Baum, S.J. (1978) Organic and Biological Chemistry. London: Macmillan Publishing Co., Inc.

Bayes, T. (1763). Philosophical Transactions of the Royal Society of London : 330-418.

Belli, M., D.T. Goodhead, F. Ianzini, T.J. Jenner, G. Simone, and M.A. Tabocchini (1991) The use of DNA precipitation assay for evaluating DSB induced by high and low LET radiations: comparison with sedimentation results. In E.M. Fielden and P. O'Neill (eds): The early effects of radiation on DNA. Berlin: Springer-Verlag, pp. 309-310.

Bengtsson, L.G. (1969) Assessment of dose equivalent from fluctuations of energy deposition. In H.G.Ebert (ed): Second Symposion on Microdosimetry. Brussels: Commission of the European Comm., pp. 375-400.

Bengtsson, L.G. (1972) Dose average event size determination for 60-Co using precision measurements of ionization current. In H.G. Ebert (ed): Third Symposium on Microdosimetry. Luxembourg: Commission of the European Communities, pp. 483-494.

Bengtsson, L.G., and L. Lindborg (1974) Comparison of pulse height analysis and variance measurements for the determination of dose mean specific energy. In J. Booz, H.G. Ebert, R. Eickel, and A. Waker (eds): Fourth Symposium on Microdosimetry. Luxembourg: Comm. Europ. Communities, pp. 823-842.

Benjamin, P.W., C.D. Kemshall, and J. Redfearn (1968) A high resolution spherical proportional counter. Nucl. Instrum. Methods 59: 77.

Bethe, H. (1930) Zur Theorie der Durchgangs schneller Korpuskularstrahlen durch Materie. Ann. Physik 5: 325.

Bethe, H. (1932) Bremsformel für Elektronen relativistischer Geschwindigkeit. Z. Phys. 76: 293.

Bethe, H. (1933) Quantenmechanik der Ein- und Zweielektronenprobleme. In H. Geiger and K. Scheel (eds): Handbuch der Physik. Berlin: Springer-Verlag, p. 273.

Bethe, H. (1953) Moliere's theory of multiple scattering. Phys. Rev. 89: 1256-1266.

Bethe, H., and W. Heitler (1934) On the stopping of fast particles and on the creation of positive electrons. Proc. Royal Soc. A146: 83.

Bird, R.P. (1979) Biophysical studies with spatially correlated ions.--3.--Cell survival studies using diatomic deuterium. Radiat. Res. 78: 210-233.

Bird, R.P., N. Rohrig, R.D. Colvett, C.R. Geard, and S.A. Marino (1980) Inactivation of synchronized Chinese hamster V-79 cells with charged particle track segments. Radiat. Res. 82: 277-289.

Boag, J.W. (1954) The distribution of linear energy as "ion density" for fast neutrons in water. Radiat. Res. 1: 323.

Bond, V.P., and M.N. Varma (1982) Low level radiation response explained in terms of fluence and cell critical volume dose. In J.Booz and H.G.Ebert (ed): Proceedings of the Eight Symposium on Microdosimetry. London: Harwood Academic, pp. 423-437.

Booz, J. (1976) Microdosimetric spectra and parameters of low LET radiation. In J. Booz, H.G. Ebert, and B.G.R. Smith (eds): Fifth Symposium on Microdosimetry. Luxembourg: Commission of European Communities, pp. 311-345.

Booz, J. (1984) Advantages of introducing microdosimetric instruments and methods into radiation protection. Microdosimetric Counters in Radiation Protection Radiat. Prot. Dos. 9: 175-183.

Booz, J., and J. Fidorra (1981) Microdosimetric investigations on collimated fast neutron beams for radiation therapy: II. The problem of radiation therapy and RBE. Phys. Med. Biol. 26: 43.

Booz, J., P. Olko, T. Schmitz, L. Morstin, and L.E. Feinendegen (1989) The KFA Counter, its Photon and Neutron Responses and its Potential for Future Developments. Rad. Prot. Dos. 29/1-2: 87-92.

Braby, L.A., G.D. Badhwar, T.J. Conroy, D.C. Elegy, and L.W. Brackenbush (1994) Automated Systems for Measuring Dose and Radiation Quality as a Function of Time. Radiat. Prot. Dos. 52: 423-426.

Braby, L.A., and W.H. Ellett (1971) Ionizations in microscopic volumes irradiated by energetic photons. Corvallis: Rad. Center, Oregon State U.

Braby, L.A., and W.C. Roesch (1980) Microdosimetry of 0.5 to 2.0 MeV Electron Beams. J.Booz, H.G.Ebert and H.D.Hartfiel (eds): Seventh Symposium on Microdosimetry p.665-676.

Bracewell, R.N. (1978) The Fourier transform and its applications. New York: McGraw Hill Book Company.

Brackenbush, L.W., and G.W.R. Endres (1985) Personnel monitors utilizing tissue equivalent proportional counters. In H. Schraube and G. Burger (eds): Fifth Symposium on Neutron Dosimetry. Luxembourg: Commission of the European Communities, pp. 359-368.

Bradford, J.N. (1980) Single event error generation by 14 MeV neutrons reactions in silicon. IEEE Trans. Nucl. Sci. NS-27: 1480-1484.

Bromberg, J.P. (1975) Measurement of absolute cross sections of electron elastically scattered by gases. In J.S. Risley and R. Geballe (eds): The physics of electronic and atomic collisions. Seattle: Univ. of Washington Press, pp. 98-111.

Bruche, E. (1929) Wirkungsquerschnitt und Molekelbau in der Pseudoedelgasreihe: Ne, HF, H_2O, NH_3, CH_4. Ann. Physik 1: 93-134.

Bueche, G., and G. Przybilla (1981) Distributions of absorbed dose from pi meson beams calculated from a new Monte Carlo program. Nucl. Instr. Meth. 179: 321.

Campion, P.J. (1972) Some Comments on the Operation of Proportional Counters. H.G.Ebert (ed): Third Symposium of Microdosimetry EURATOM EUR 4810 d-f-e : 601-611.

Caswell, R.S., and J.J. Coyne (1974) Neutron energy deposition spectra studies. In J. Booz, H.G. Ebert, R. Eickel, and W. Walker (eds): Fourth Symposium on Microdosimetry. Verbania-Pallanza. Luxembourg: Commission of the European Comm., p. 967.

Cauchy, A. (1908) Memoire sur la rectification des courbes et la quadrature des sourface courbe. In (ed): Oevres Completes. Paris: Gauthier Villard.

Chandrasekhar, S. (1943) Stochastic problems in physics and astronomy. Rev. Mod. Phys. 15: 1-88.

Chatterjee, A., and H.J. Schaefer (1976) Microdosimetric structure of heavy ion tracks in tissue. Radiat. Environ. Biophys. 13: 215-227.

Chen, J., J. Breckow, H. Roos, and A.M. Kellerer (1990) Further development of the variance-covariance method. Rad. Prot. Dos. 31: 171-173.

Chmelevsky, D., A.M. Kellerer, M. Terrissol, and J.P. Patau (1980) Proximity functions for electrons up to 10 keV. Radiat. Res. 84: 219-38.

Clifford, P., N.J.B. Green, and M.J. Pilling (1982) Monte Carlo simulation of diffusion and reaction in radiation-induced spurs. Comparison with analytic models. J. Phys. Chem. 86: 1322-1327.

Clifford, P., N.J.B. Green, and M.J. Pilling (1982) Stochastic model based on pair distribution functions for reaction in a radiation-induced spur containing one type of radical. J. Phys. Chem. 86: 1318-1321.

Coleman, R. (1973) Random paths through rectangles and cubes. Metallography 6: 103-114.

Coppola, M., and J. Booz (1975) Neutron scattering and energy deposition spectra. Radiat. Environ. Biophys. 12: 157.

Curtis, S.B. (1986) Lethal and potentially lethal lesions induced by radiation: a unified repair model. Radiat. Res. 106: 252.

Debye, P. (1942) Reaction rates in ionic solutions. Trans. Electrochem. Soc. 82: 265-272.

Dicello, J.F. (1987) Microelectronics and microdosimetry. Nucl. Instr. Meth. B24/25: 1044-1049.

Dicello, J.F., W. Gross, and U. Kraljevic (1972) Radiation quality of californium-252. Phys. Med. Biol. 17: 345.

Dicello, J.F., M. Zaider, and D.J. Brenner (1980) Dosimetry of pions. In R.H. Thomas and V. Perez-Mendez (eds): Advances in Radiation Protection and Dosimetry in Medicine. N.Y.: Plenum Press.

Edwards, A.A., and D.C. Lloyd (1991) Chromosomal damage in human lymphocytes: effect of radiation quality. In E.M. Fielden and P. O'Neill (eds): The early effects of radiation on DNA. Berlin: Springer-Verlag, pp. 385-396.

Ehrenreich, H., and M.H. Cohen (1959) Self-consistent field approach to the many-electron problem. Phys. Rev. 115: 786-790.

Ellet, W.H., and L.A. Braby (1972) Measurement of Event Spectra from 250 kVp x-rays Cobalt 60 Gamma Rays and Tritium Beta Rays using a Grid Walled Proportional Counter. H.G.Ebert (ed.): Third Symposium on Microdosimetry. EURATOM Document EUR 4810 d-f-e : 471-481.

EURADOS (1984) Eurados Symposium on "Microdosimetric Counters in Radiation Protection. Rad. Prot. Dos. 9:.

EURADOS (1989) Eurados Symposium on "Implementation of Dose-Equivalent Meters Based on Microdosimetric Techniques in Radiation Protection". Rad. Prot. Dos. 29/1-2:.

Evans, R.D. (1955) The Atomic Nucleus. New York: McGraw-Hill Book Company, Inc.

FAA (1975) Cosmic Radiation Exposure in Supersonic and Subsonic Flight. Report to the Federal Aviation Administration. Aviat. Space and Env. Med. : 1170-1185.

Fano, U. (1947) Ionization Yield of Radiations. II. The Fluctuations of the Number of Ions. Phys. Rev. 42: 26-29.

Fano, U. (1954) Note on the Bragg-Gray Cavity Principle for Measuring Energy Dissipation. Rad. Res. 1: 237-240.

Fermi, E. (1950) Nuclear Physics University of Chicago Press.

Forsberg, B., M. Jensen, L. Lindborg, and G. Samuelson (1978) Determination of the dose mean of specific energy for conventional x-rays by variance measurements. In J. Booz and H.G. Ebert (eds): Sixth Symposium on Microdosimetry. Brussels: Harwood Academic Publishers Ltd., pp. 261-271.

Glass, W.A., and L.A. Braby (1969) A Wall-less Detector for Measuring Energy Deposition Spectra. Radiat. Res. 39: 230-240.

Goldberger, M.L., and K.M. Watson (1964) Collision Theory. New York: John Wiley & Sons.

Goldhagen, P., and G. Randers-Pehrson (1992) Variance-covariance: a practical method for microdosimetry in submicroscopic volumes. In D.C. Dewey, M. Edington, R.J.M. Fry, E.J. Hall, and G.F. Whitmore (eds): Radiation Research: A Twentieth Century Perspective. San Diego: Academic Press, p. 415.

Goldhagen, P., G. Randers-Pehrson, S.A. Marino, and P. Kliauga (1990) Variance-covariance measurements of y_D for 15 MeV neutrons in a wide range of site sizes. Rad. Prot. Dos. 31: 167-170.

Goodman, L.J. (1969) A modified tissue-equivalent liquid. Health Phys. 16 : 763.

Goodman, L.J. (1979) Evaluation of dose equivalent using a tissue-equivalent ionization chamber and a Geiger-Muller dosimeter. Radiat. and Environmental Biophys. 16: 367-371.

Goodman, L.J., and J.J. Coyne (1980) Wn and neutron kerma for methane-based tissue equivalent gas. Radiat. Res. 82: 13-26.

Green, A.E.S., and T. Sawada (1972) Ionization cross sections and secondary electron distributions. J. Atmos. Terr. Phys. 34: 1719-1728.

Green, A.E.S., and P.S. Stolarski (1972) Analytic models of electron impact excitation cross sections. J. Atmos. Terr. Phys. 34: 1703-171.

Green, S., A.C.A. Aro, G.C. Taylor, and M.C. Scott (1990) The Development of Microdosimetric Detectors for Investigating LET Distributions in Different Body Tissues. Rad. Prot. Dos. 31/1-4: 137-141.

Gross, W., and R.C. Rodgers (1972) Heavy ion event spectra. H.G.Ebert (ed.): Third Symposium on Microdosimetry : 873-887.

Grosswendt, B., and E. Waibel (1978) Transport of low energy electrons in nitrogen and air. Nucl. Instrum. Methods 155: 145-156.

Guenther, K., W. Schultz, and W. Leistner (1977) Microdosimetric approach to cell survival curves in dependence on radiation quality. Stud. Biophys. 61: 163-209.

Gull, S.F., and G.J. Daniell (1978) Image reconstruction from incomplete and noisy data. Nature 272: 686-690.

Hall, E.J., W. Gross, R.F. Dvorak, A.M. Kellerer, and H.H. Rossi (1972) Survival curves and age response functions for Chinese hamster cells exposed to x rays or high LET alpha particles. Radiat. Res. 52: 88-98.

Hamm, R.N., J.E. Turner, H.A. Wright, and R.H. Ritchie (1981) Calculated distance distributions of energy transfer events in irradiated water. In J. Booz, H.G. Ebert, and H.D. Hartfiel (eds): Seventh Symposium on

Microdosimetry, Oxford. Harwood, London: Comm. of the European Communities, pp. 717-726.

Harder, D., and P. Virsik (1987) Kinetics of cell survival as predicted by the repair-interaction model. In: Proceedings of the 8-th Congress of Radiation Research. New York: Taylor and Francis, p. 318.

Hartmann, G., H.G. Menzel, and H. Schulmacher (1981) Different Approaches to Determine Effective Quality Factors and Dose Equivalent Using the Rossi Counter. G.Burger and H.G.Ebert (eds): Fourth Symposium on Neutron Dosimetry. Commission of the European Communities EUR 7448 EN.

Hedin, L., and S. Lundqvist (1969) Effects of electron-electron and electron-phonon interactions on the one-electron states of solids. Solid State Phys $\underline{23}$: 1-181.

Hei, T.K., K. Komatsu, E.J. Hall, and M. Zaider (1988) Oncogenic transformation by charged particles of defined LET. Carcinogenesis $\underline{7}$: 747-750.

Heller, J.M., R.N. Hamm, R.D. Birkhoff, and L.R. Painter (1974) Collective oscillation in liquid water. J. Chem. Phys. $\underline{60}$: 3483-3486.

Horowitz, Y.S. (1981) Theoretical and microdosimetric basis of thermoluminescence and applications to dosimetry. Phys. Med. Biol. $\underline{26}$: 765-824.

Hubbell, J.H. (1982) Photon mass attenuation and energy absorption coefficients from 1 keV to 20 MeV. Int. J. Appl. Radiat. Isot. $\underline{33}$: 1269.

Hug, O., and A.M. Kellerer (1966) Stochastik der Strahlenwirkung, Springer Verlag Berlin.

Hutchinson, F. (1954) Energy Requirements for the Inactivation of Bovine Serum Albumin by Radiation. Rad. Res. $\underline{1}$: 43-52.

ICRP (1955) Report of the International Commission on Radiological Protection. Brit. Journ. Radiology Supplement 6.

ICRP (1991) Recommendations of the International Commission on Radiological Protection. Annals of the ICRP $\underline{21}$.

ICRU (1962) Radiation Quantities and Units. Report 10a. International Commission on Radiological Units and Measurements. Bethesda, MD.

ICRU (1970) Lineal energy transfer, ICRU Report 16. Washington, DC.

ICRU (1977) Neutron Dosimetry for Biology and Medicine, ICRU Report 26. Bethesda, MD: Int. Comm. Radiat. Units. Meas.

ICRU (1978) Basic aspects of high energy particle interactions and radiation dosimetry. ICRU Report 28. Bethesda, MD.

ICRU (1979) Average Energy Required to Produce an Ion Pair. ICRU Report 39. Bethesda, MD.

ICRU (1980) Radiation Quantities and Units, ICRU Report 33. Bethesda, MD: Int. Comm. Radiat. Units and Meas.

ICRU (1983) Microdosimetry. ICRU Report 36. Bethesda, MD.

ICRU (1984) Stopping powers for electrons and positrons. ICRU Report 37. Bethesda, MD.

ICRU (1985) Determination of Dose Equivalents Resulting from External Radiation Sources, ICRU Report 39. Bethesda, MD.

ICRU (1986) The quality factor in radiation protection. International Commission on Radiation Units and Measurements, ICRU Report 40. Bethesda, MD.

ICRU (1988) Determination of Dose Equivalents from External Radiation Sources - Part 2, ICRU Report 43. Bethesda, MD.

ICRU (1992) Quantities and Units in Radiation Protection Dosimetry. ICRU Report 51. Bethesda, MD.

ICRU (1993) Measurement of Dose Equivalents from External Photon and Electron Radiations. ICRU Report 47. Bethesda, MD.

Inokuti, M. (1971) Inelastic collisions of fast charged particles with atoms and molecules - the Bethe theory revisited. Rev. Modern. Phys. 43: 297-347.

Jackson, J.D. (1975) Classical Electrodynamics. New York: John Wiley & Sons.

Jaynes, E.T. (1957) Information theory and statistical mechanics. Phys. Rev. 106: 620-630.

Johns, H.E., and J.R. Cunningham (1983) The Physics of Radiology. Springfield, Ill: Charles C. Thomas.

Kahn, H. (1956) Applications of Monte Carlo. Santa Monica, CA: The Rand Corporation.

Kampf, G. (1988) Induction of DNA double-strand breaks by ionizing radiation of different quality and their relevance for cell inactivation. Radiobiol. Radiotherapy 29: 631-658.

Kellerer, A.M. (1968) Mikrodosimetrie, Grundlagen einer Theorie der Strahlungsqualität. Gesellschaft fur Strahlenforschung GSF Ser. Monogr. B1 Munich.

Kellerer, A.M. (1970) Analysis of Patterns of Energy Deposition EURATOM Report EUR 4452 d-e-f. Brussels. H.G.Ebert (ed): Second Symposium on Microdosimetry : 107-134.

Kellerer, A.M. (1971) An assessment of Wall Effects in Microdosimetric Measurements. Radiat. Res. 47: 377-386.

Kellerer, A.M. (1972) An algorithm for LET analysis. Phys. Med. Biol. 17: 232.

Kellerer, A.M. (1980) Concepts of geometrical probability relevant to microdosimetry and dosimetry. In J. Booz, H.G. Ebert, and H.D. Hartfiel (eds): Seventh Symposium on Microdosimetry. Oxford, UK: Harwood, London, pp. 1049-1062.

Kellerer, A.M. (1981) Concepts of geometrical probability relevant to microdosimetry. In J. Booz, H.G. Ebert, and H.D. Hartfiel (eds): Seventh Symposium on Microdosimetry. EUR 7147 de-en-fr. Harwood, London: Academic Publishers, pp. 1049-62.

Kellerer, A.M. (1984) Chord length distributions and related quantities for spheroids. Radiat. Res. 98: 425-437.

Kellerer, A.M. (1985) Fundamentals of microdosimetry. In K.R. Kase, B.E. Bjarngard, and F.H. Attix (eds): The dosimetry of ionizing radiation. New York: Academic Press, Inc.

Kellerer, A.M., and D. Chmelevsky (1975) Concepts of microdosimetry III mean values of the microdosimetric distributions. Rad. and Environ. Biophys. 12: 321-335.

Kellerer, A.M., and K. Hahn (1988) Considerations on a Revision of the Quality Factor. Rad. Res. 114: 480-488.

Kellerer, A.M., and O. Hug (1972) Theory of dose-effect relations. In A. Zuppinger and O. Hug (eds): Encyclopedia of Medical Radiology. Heidelberg: Springer-Verlag, pp. 1-42.

Kellerer, A.M., Y.P. Lam, and H.H. Rossi (1980) Biophysical studies with spatially correlated ions. 4. Analysis of cell survival data for diatomic deuterium. Radiat. Res. 83: 522-528.

Kellerer, A.M., and H.H. Rossi (1972) The theory of dual radiation action. Curr. Top. Radiat. Res. Q 8: 85-158.

Kellerer, A.M., and H.H. Rossi (1978) A generalized formulation of dual radiation action. Radiat. Res. 75: 471-88.

Kellerer, A.M., and H.H. Rossi (1984) On the determination of microdosimetric parameters in time-varying radiation fields: The variance-covariance method. Radiat. Res. 97: 237-245.

Kellerer, A.M., H.H. Rossi, M. Lassman, and A. Weuning (1984) Microdosimetric Computations for Heavy Ion Beams. GSI Scientific Report 250.

Kendall, M.G., and P.A.P. Moran (1963) Geometrical Probability. London: Griffin.

Kim, Y.S. (1973) Density Effect in dE/dx of Fast Charged Particles Traversing various Biological Materials. Radiat. Res. : 21-27.

Kissel, A., C.A. Quarles, and R.H. Pratt (1983) Shape functions for atomic-field bremsstrahlung from electrons of kinetic energy 1-500 keV in selected neutral atoms (1<Z<92). Atomic Data and Nuclear Data Tables 28: 381.

Kliauga, P.J. (1990) Measurement of Single Event Energy Deposition Spectra at 5 μm to 250 nm Simulated Site Sizes. Rad. Prot. Dos. 31/3-4: 119-123.

Kliauga, P.J., R.T. Colvett, L.J. Goodman, and Y.M. Lam (1978) Microdosimetry of 400 MeV/AMU 12C and 40Ar Beams. J.Booz and H.G.Ebert (eds): Sixth Symposium on Microdosimetry Harwood, London : 1173-1183.

Kliauga, P.J., R.D. Colvett, M.S.Y.P. Lam, and H.H. Rossi (1978) The relative biological effectiveness of 160 MeV protons. I. Microdosimetry. Int. J. Radiat. Oncology Biol. Phys. 4: 1001-1008.

Kliauga, P.J., and R. Dvorak (1978) Microdosimetric measurements of ionization by monoenergetic photons. Radiat. Res. 73: 1-20.

Kliauga, P.J., H.H. Rossi, and G. Johnson (1989) A multi-element counter for radiation protection measurement. Health Phys. 57/4: 631.

Kuehner, A.V., J.D. Chester, and J.W. Baum (1973) A Portable Mixed Radiation Dose Equivalent Meter. International Atomic Energy Agency IAEA-SM-1 67/50: 233-246.

Kunz, A., E. Pihet, E. Arnaud, and H.G. Menzel (1990) An Easy-to-Operate Pulse Height Analysis System for Area Monitoring with TEPC in Radiation Protection. Nucl. Inst. and Meth. in Phys. Res. A 299: 696-701.

Kuppermann, A. (1961) Diffusion kinetics in radiation chemistry. In: The chemical and biological action of radiations. NY: Academic Press, pp. 87-166.

Landau, L. (1944) On the energy loss of fast charged particles. J. Phys. USSR $\underline{8}$: 210-205.

Li, K., P. Kliauga, and H.H. Rossi (1984) Microdosimetry and thermoluminescence. Radiat. Res. $\underline{99}$: 465-475.

Lindborg, L. (1974) Microdosimetry in high energy electron and ^{60}Co gamma ray beams for radiation therapy. In: J.Booz, H.G.Ebert, R.Eickel, and W.Walker (eds): Fourth Symposium on Microdosimetry : 799-817.

Lindborg, L. (1976) Microdosimetric measurements in beam of high energy photons and electrons: techniques and results. In J. Booz, H.G. Ebert, and B.G.R. Smith (eds): Fifth Symposium on Microdosimetry. Luxembourg: Commission of European Communities, p. 347.

Lindborg, L., P. Kliauga, S. Marino, and H.H. Rossi (1985) Variance-Covariance measurements of the dose mean lineal energy in a neutron beam. Radiat. Prot.Dos. $\underline{13(1-4)}$: 347-351.

Lindborg, L.R. (1974) Microdosimetry in high energy electron and 60Co gamma ray beams for radiation therapy. In J. Booz, H.G. Ebert, R. Eickel, and W. Walker (eds): Fourth Symposium on Microdosimetry, Verbania-Pallanza. Luxembourg: Commission of the European Comm., pp. 24-28.

Lloyd, D.C., and A.A. Edwards (1983) Chromosome aberrations in human lymphocytes: effect of radiation quality, dose and dose rate. In: Radiation-induced chromosome damage in man. New York: Alan R. Liss, Inc, pp. 23-49.

Lloyd, D.C., R.J. Purrott, G.W. Dolphin, and A.A. Edwards (1975) Relationship between chromosome aberrations and low LET radiation dose to human lymphocytes. Int. J. Radiat Biol. $\underline{28}$: 75-90.

Lloyd, E.L., M.A. Gemmell, C.B. Henning, D.S. Gemmell, and B.J. Zabransky (1979) Cell survival following multiple-track alpha particle irradiation. Int. J. Radiat. Biol. $\underline{35}$: 23-31.

Luxton, G., P. Fessenden, and W. Hoffmann (1979) Microdosimetric Measurements of Pretherapeutic Heavy Ion Beams. Radiat. Res. $\underline{79}$: 256-272.

Mahan, G.D. (1990) Many-particle physics. New York: Plenum Press.

Mark, T.D., and F. Egger (1976) Cross section for single ionization of H_2O and D_2O by electron impact from threshold up to 170 eV. Int. J. Mass Spectrom. Ion Phys. $\underline{20}$: 89-99.

McNulty, P.J. (1990) Predicting single event phenomena. In: IEEE 1990 International Nuclear and Space Radiation Effects Conference. Reno, Nevada:, pp. 3/1-3/93.

McQuarrie, D.A. (1967) Stochastic approach to chemical kinetics. J. Appl. Prob. $\underline{4}$: 413-478.

Menzel, H.G. (1984) Proportional Counter Measurements in Neutron Therapy Beams. In "Advances in Dosimetry for Fast Neutrons and Heavy Charged Particles for Therapy Applications". IAEA, Vienna 371/1.

Menzel, H.G., G. Buelher, and H. Schuhmacher (1983) Investigation of basic uncertainties in the experimental determination of microdosimetric data. In J. Booz and H.G. Ebert (eds): Eight Symposium on Microdosimetry. Luxembourg: Commission of the European Communities, pp. 1061-1072.

Messel, H., and D.F. Crawford (1970) Electron-photon shower distribution function. Tables for lead, copper and air absorbers. New York: Pergamon Press.

Metting, N.F., H.H. Rossi, L.A. Braby, P.J. Kliauga, J. Howard, M. Zaider, W. Schimmerling, and M. Wong (1988) Microdosimetry near the trajectory of high-energy heavy ions. Radiat. Res. 116: 183-195.

Mills, R.E., and H.H. Rossi (1980) Mean energy deposition distribution about proton tracks. Radiat. Res. 84: 434-443.

Moeller, C. (1932) Theorie der Durchgangs schneller Elektronen durch Materie. Ann. Physik 14: 531.

Moliere, G. (1947) Theory of scattering by fast charged particles. I. Single scattering in a screened Coulomb field. Z. Naturforsch. 2a: 133.

Morse, P.M., and H. Feshbach (1953) Methods of Theoretical Physics. New York: McGraw Hill Publishing Co.

NCRP (1991) Conceptual basis for calculations of absorbed-dose distributions. Bethesda, MD: National Council for Radiation Protection and Measurements.

Nguyen, V.D., P. Bouisset, Y.A. Akatov, V.M. Petrov, S.B. Kozlova, M. Siegrist, and J.F. Szilling (1990) Measurements of Quality Factors and Dose Equivalents with CIRCE Inside the Soviet Space Station MIR. Rad. Prot. Dos. 31,1/4: 377-382.

Nguyen, V.D., M. Chemtob, B. Lavigne, and N. Parmentier (1976) Etude Microdosimetrique d'un Faisceau d'Helious de 649 MeV. J.Booz, H.G.Ebert, and B.G.R.Smith (eds): Fifth Symposium on Microdosimetry Euratom Report EUR 5452, Luxembourg : 153-166.

Nishimura, H. (1979) Elastic scattering cross sections of H2O by low energy electrons. In K. Takayanagi and N. Oda (eds): Electronic and Atomic Collisions. Kyoto: Society for Atomic Coll. Res., p. 314.

Obelic, B. (1985) Probability Density Distribution of Primary Ionization Calculated by Means of Iterative Deconvolution Process. Nucl. Instr. and Meth. in Phys. Res. A241: 515-518.

Paretzke, H.G., G. Leuthold, G. Burger, and W. Jacobi (1974) Approaches to physical track structure calculation. In J. Booz, H.G. Ebert, R. Eickel, and A. Waker (eds): Fourth Symposium on Microdosimetry. Luxembourg: Commission of Eur. Communities, pp. 123-140.

Pihet, P., H.G. Menzel, W.G. Alberts, and H. Kluge (1989) Response of Tissue-Equivalent Proportional Counters to Low and Intermediate Energy Neutrons Using Modified TE-3He Gas Mixtures. Rad. Prot. Dos. 29/1-2: 113-118.

Pihet, P., H.G. Menzel, R. Schmidt, M. Beauduin, and A. Wambersie (1990) Biological Weighting Function for RBE Specification of Neutron Therapy Beams. Intercomparison of 9 European Centers. Rad. Prot. Dos. $\underline{31,1/4}$: 437-442.

Pines, D., and P. Nozieres (1966) The theory of quantum liquids. New York: Benjamin.

Porter, H.S., and F.W. Jump (1978) Analytic total and angular elastic electron impact cross sections for planetary atmospheres. Computer Science Corp. Report CSC/TM-78/6017 .

Powell, S., and T.J. McMillan (1990) DNA damage and repair following treatment with ionizing radiation. Radiotherapy and Oncology $\underline{19}$: 95-108.

Powers, E.L., J.T. Lyman, and C.A. Tobias (1968). Some effects of accelerated charged particles on bacterial spores. Int. J. Radiat. Biol. $\underline{14}$: 313.

Pszona (1976) A Track Ion Counter. J.Booz, H.G.Ebert, and B.G.R.Smith (eds): Fifth Symposium on Microdosimetry EURATOM Report EUR 5452 d-e-f : 1107-1121.

Radeka, V. (1968) State of the art of low noise amplifiers for semiconductor radiation detectors. In (ed): Proceedings of International Symposium in Nuclear Electronics. Springfield, VA: National Technical Information Service, p.

Randers-Pehrson, G., R.W. Finlay, J.F. Dicello, and J.C. McDonald (1983) A technique for time-resolved microdosimetric spectroscopy. In J. Booz and H. Ebert (eds): Radiation Protection. Eights Symposium on Microdosimetry. Julich, Germany:, pp. 1169-1177.

Ritchie, R.H., R.N. Hamm, J.E. Turner, H.A. Wright, J.C. Ashley, and G.J. Basbas (1988) Physical aspects of charged particle structure. Nucl. Tracks. Radiat. Measur. $\underline{16}$: 141-158.

Robbins, H.E. (1944) On the measure of a random set II. Ann. Math. Stat. $\underline{16}$: 342-347.

Rodgers, R.C., J.F. Dicello, and W. Gross (1973) The biophysical properties of 3.9 GeV nitrogen ions. II Microdosimetry. Radiat. Res. $\underline{54}$: 12-23.

Roesch, W.C. (1967) A model for the action of radiation on simple biological systems. In: Proc. 1-st Int. Symp. on the Biol. Interact. of Dose from Acid. Prod. Rad. USAEC Washington Conf. 670305. , p. 297.

Roesch, W.C. (1968) Mathematical theory of radiation fields. In F.H. Attix and W.C. Roesch (eds): Radiation Dosimetry. New York: Academic Press.

Roots, R., and S. Okada (1975) Estimation of life times and diffusion distances of radicals involved in x-ray-induced DNA strand breaks of killing of mammalian cells. Radiat. Res. $\underline{64}$: 306-20.

Rosenzweig, W., and H.H. Rossi (1959) Determination of the quality of the absorbed dose delivered by monoenergetic neutrons. Rad. Res. $\underline{10(5)}$: 532-544.

Rossi, B. (1952) High-energy particles. New York: Prentice-Hall, Inc.

Rossi, H.H. (1967a) Energy deposition in the absorption of radiation. Adv. Biol. Med. Phys. $\underline{11}$: 27-85.

Rossi, H.H. (1967b) Microscopic energy distribution in irradiated matter. In: Radiation Dosimetry. NY: Academic Press, pp. 43-92.

Rossi, H.H. (1970) The Effects of Small Doses of Ionizing Radiation. Radiation Phys. Med. Biol. 15: 255-262.

Rossi, H.H. (1979a) Biophysical studies with spatially correlated ions.1. Background and theoretical considerations. Radiat. Res. 78: 185-191.

Rossi, H.H. (1979b) The RBE of neutrons. In: Proc. of a Conf. on Neutrons from Electron Medical Accelerators. Gaithersburg,MD: NBS, pp. 37-39.

Rossi, H.H. (1979c) The role of microdosimetry in radiobiology. Radiat. Environmental Biophys. 17: 29-40.

Rossi, H.H. (1984) Limitation and Assessment in Radiation Protection. L.S. Taylor Lecture No. 8. National Council on Radiation Protection and Measurement, Bethesda, MD.

Rossi, H.H. (1994) Geometric Domains in Cellular Radiobiology. Radiat. Prot. Dos. 52: 9-12.

Rossi, H.H., M.W. Biavati, and W. Gross (1961) Local energy density in irradiated tissues I Radiobiological Significance. Radiat. Res. 15/4: 431-439.

Rossi, H.H., and G. Failla (1956) Neutrons dosimetry. In O. Glasser (ed): Medical Physics. Chicago: Year Book Publishers, Inc., pp. 603-607.

Rossi, H.H., and A.M. Kellerer (1972) Radiation Carcinogenesis at Low Doses. Science 175: 200-202.

Rossi, H.H., and A.M. Kellerer (1973) Biological implications of microdosimetry: I. Temporal aspects. In J. Booz, H.G. Ebert, R. Eickel, and A. Waker (eds): Fourth Symposium on Microdosimetry. Luxembourg: CEC, pp. 315-330.

Rossi, H.H., and A.M. Kellerer (1986) The dose rate dependence of oncogenic transformation by neutrons may be due to variation of response during the cell cycle. Int. J. Radiat. Biol. 50: 353-361.

Rossi, H.H., and W. Rosenzweig (1955) Measurements of neutron dose as a function of linear energy transfer. Rad. Res. 2: 417-425.

Rossi, H.H., W. Rosenzweig, M.H. Biavati, L. Goodman, and L. Phillips (1962) Radiation Protection Surveys at Heavy-Particle Accelerators Operating at Energies Beyond Several Hundred Million Electron-Volts. Health Phys. 8: 331-342.

Rossi, H.H., and M. Zaider (1992) Compound dual radiation action. I. General aspects. Radiat. Res. 132: 178-183.

Samuel, A.H., and J.L. Magee (1953) Theory of radiation chemistry. II. Track effects in radiolysis of water. J. Chem. Phys. 21: 1080-1087.

Santalo, L.A. (1976) Integral geometry and geometric probability. Reading, MA: Addison-Wesley Publishing Co.

Savage, J.R.K. (1975) Classification and relationships of induced chromosomal structural changes. J. Medical Genetics 12: 103-122.

Schattschneider, P. (1986) Fundamentals of Inelastic Electron Scattering. New York: Springer Verlag.

Schmitz, T.H., and J. Booz (1989) Measurement of the Gas Amplification Coefficient in a TEPC. Rad. Prot. Dos. 29/1-2: 31-36.

Schuhmacher, H., and O. Krauss (1986) Area Monitoring of Photons and Neutrons from Medical Electron Accelerators Using Tissue-Equivalent Proportional Counters. Radiat. Prot. Dos. 14/4: 325-327.

Schuhmacher, H., H.G. Menzel, H. Blattmann, and H. Muth (1985) Proportional counter dosimetry and microdosimetry for radiotherapy with multiple pion beams. Radiat. Res. 101: 177.

Schutten, J., F.J. de Heer, H.R. Moustafa, A.J.H. Boerboom, and J. Kistemaker (1966) Gross- and partial-ionization cross sections for electrons on water vapor in the energy range 0.1-20 keV. J. Chem. Phys. 44: 3924-3928.

Segur, P. Peres, S. Boeuf, and J. Barthe (1990) Modeling of the Electron and Ion kinetics in Cylindrical Proportional Counters. Rad. Prot. Dos. 31/1-4: 107-118.

Seng, G., and F. Linder (1976) Vibrational excitation of polar molecules by electron impact. II. Direct and resonant excitation in $H2O$. J. Phys. B9: 2539-2551.

Setlow, R. (1960) Ultraviolet Wave-length-dependent Effects on Proteins and Nucleic Acids. Rad. Res. Suppl.2: 276-289.

Shannon, C.E. (1948) A mathematical theory of communication. Bell System Technical Journal 27: 379-423.

Shonka, R.F., J.E. Rose, and G. Failla (1958) Conducting Plastic Equivalent to Tissue, Ion and Polystyrene. 2nd United Nations International Conference on Peaceful Uses of Atomic Energy 21: 184.

Simpson, J.A. (1983) Introduction to the galactic cosmic radiation. In M.M. Shapiro (ed): Composition and Origin of Cosmic Rays. Dorcht, Netherlands: Reidel Publishing Co., p. 1.

Skarsgard, L.K., A. Kihlman, L. Parker, C.M. Pujara, and S. Richardson (1967) Survival, chromosome abnormalities recovery in heavy ion- and X-irradiated mammalian cells. Radiat. Res.[Suppl.] 7: 208-221.

Srdoc, D. (1970) Experimental Technique of Measurement of Microscopic Energy Distribution in Irradiated Matter using Rossi Counters. Rad. Res. 43: 302-319.

Srdoc, D., and B.C. Clark (1970) Generation and Spectroscopy of Ultrasoft x-Rays by Non-Dispersive Methods. Nucl. Instr. & Meth. 69: 1-9.

Srdoc, D., L.J. Goodman, S.A. Marino, R.E. Mills, M. Zaider, and H.H. Rossi (1981) Microdosimetry of monoenergetic neutron radiation. In J. Booz, H. Ebert, and H. Hartfiel (eds): Seventh Symposium on Microdosimetry. London: Harwood Academic Publishers, pp. 765-774.

Srdoc, D., M. Inokuti Krajar-Bronic (1993) Yields of Ionization and Excitation in Irradiated Matter. IAEA CRP Atomic and Molecular Data for Radiotherapy.

Srdoc, D., B. Obelic, and I. Kajar Bronie (1987) Statistical fluctuations in the ionization yield for low-energy photons absorbed in polyatomic gases. J. Phys. B: Atm. Molec. Phys. 20: 4473.

Srdoc, D., and A. Sliepcevic (1963) Carbon dioxide counter: effect of gaseous impurities and gas purification method. J. Appl. Radiat. Isotopes 14: 481.

Stanton, J.A., G. Taucher-Scholz, M. Schneider, and G. Kraft (1990) Comparison between indirect and direct effects for high and low LET radiations in SV40 strand break induction. Radiat. Prot. Dos. 31: 253.

Stonell, G.P., M. Marschall, J.A. Simmons, and J.A. Track (1993) Studies in Water Vapor Using a Low Pressure Cloud Chamber I Macroscopic Measurements. II Microdosimetric Measurements. Rad. Res. 136: 341-360.

Tobias, C.A. (1985) The repair-misrepair model in radiobiology: Comparison to other models. Radiat. Res. Suppl.8: S77.

Turner, J.E., R.N. Hamm, H.A. Wright, J.T. Modolo, and G.M.A.A. Sardi (1980) Monte Carlo calculations of initial energies of Compton electrons and photoelectrons in water irradiated by photons with energies up to 2 MeV. Health Phys 39: 49-55.

Turner, J.E., J.L. Magee, H.A. Wright, A. Chatterjee, R.N. Hamm, and R.H. Ritchie (1983) Physical and chemical development of electron tracks in liquid water. Radiat. Res. 96: 437-49.

Varma, M.N., and J.W. Baum (1980) Energy deposition in nanometer regions by 377 MeV/nucleon 20-Ne ions. Radiat. Res. 81: 355.

Varma, M.N., C.S. Wuu, and M. Zaider (1994) Hit-size effectiveness in relation to the microdosimetric site size. Radiat. Prot. Dos. 52: 339-346.

Vavilov, P.V. (1957) Ionization losses of high-energy heavy particles. J. Experimental Theoret. Phys. (USSR) 32: 920-923.

von Engel, A. (1965) Ionized Gases. Clarendon Press, Oxford.

Wambersie, A., P. Pichet, and H.G. Menzel (1990) The Role of Microdosimetry in Radiotherapy. Rad. Prot. Dos 31,1/4: 421-432.

Wingate, C.L., and J.W. Baum (1976) Measured radial distribution of dose and LET for alpha and proton beams in hydrogen and tissue equivalent gas. Radiat. Res. 65: 1-19.

Wuu, T.Y. (1962) Quantum theory of scattering. Engelwood Cliff, New Jersey: Prentice Hall.

Zaider, M. (1990) Concepts for describing the interaction of two agents. Radiat. Res. 123: 257-262.

Zaider, M. (1993) A mathematical formalism describing the yield of radiation-induced single- and double-strand DNA breaks, and its dependency on radiation quality. Radiat. Res. 134: 1-8.

Zaider, M., and D.J. Brenner (1984a) The application of track calculations to radiobiology.--III. Analysis of the molecular beam experiment results. Radiat.Res. 100: 213-221.

Zaider, M., and D.J. Brenner (1984b) On the stochastic treatment of fast chemical reactions. Radiat. Res. 100: 245-256.

Zaider, M., and D.J. Brenner (1985) On the Microdosimetric Definition of Quality Factors. Radiation Research 103: 302-316.

Zaider, M., and D.J. Brenner (1986) Evaluation of a specific quality function for mutation induction in human fibroblasts. Radiat. Protec. Dos. 15: 79-82.

Zaider, M., D.J. Brenner, K. Hanson, and G.N. Minerbo (1982) An algorithm for determining the proximity distribution from dose-averaged lineal energies. Radiat. Res. 91: 95-103.

Zaider, M., D.J. Brenner, and W.E. Wilson (1983) The application of track calculations to radiobiology. I. Monte Carlo simulation of proton tracks. Radiat. Res. 95: 231-47.

Zaider, M., J.L. Fry, and D.E. Orr (1990) Towards an ab initio evaluation of the wave vector and frequency dependent dielectric response function for crystalline water. Radiat. Prot. Dosim. 31: 23-28.

Zaider, M., A.Y.C. Fung, and M. Bardash (1994) Charged-particle transport in biomolecular media: the third generation. In M.N. Varma and A. Chatterjee (eds): Computational Approaches in Molecular Biology: Monte Carlo Methods. New York: Plenum Press, p.

Zaider, M., and G.N. Minerbo (1988a) On the possibility of obtaining non-diffused proximity functions from cloud-chamber data: I. Fourier deconvolution. Phys. Med. Biol. 33: 1261-1272.

Zaider, M., and G.N. Minerbo (1988b) On the possibility of obtaining non-diffused proximity functions from cloud-chamber data: II Maximum entropy and Bayesian methods. Phys. Med. Biol. 33: 1273-1284.

Zaider, M., and H.H. Rossi (1980) The synergistic effects of different radiations. Radiat. Res. 83: 732-739.

Zaider, M., and H.H. Rossi (1985) Dual radiation action and the initial slope of survival curves. Radiat. Res. 104: 568-576.

Zaider, M., and H.H. Rossi (1988a) Indirect effects in dual radiation action. Radiat. Phys. Chem. 32: 143-148.

Zaider, M., and H.H. Rossi (1988b) On the application of microdosimetry to radiobiology. Radiat. Res. 113: 15-24.

Zaider, M., and H.H. Rossi (1989) Estimation of the quality factor on the basis of multi-event microdosimetric distributions. Health Phys 56: 885-892.

Zaider, M., and H.H. Rossi (1992) On the question of RBE reversal at high doses. Radiat. Res. 130: 117-120.

Zaider, M., and M.N. Varma (1992) Nanodosimetry of radon alpha particles. In F.T. Cross (ed): Proceedings of the 29-th Hanford Symposium on Health and Environment. Columbus, Richland: Battelle Press, pp. 291-306.

Zaider, M., and C.S. Wuu (1995) The survival probability of cells exposed, at low dose rate, to radioactive sources: the effects of sublethal damage repair and cell-cycle progression. British J. Radiol. 68: 58-63.

Zaider, M., and C.Z. Wuu (1994) The combined effects of sublethal damage repair, cellular repopulation and redistribution in the mitotic cycle. I. Survival probabilities following exposure to radiation. Radiat. Res. (submitted):.

Zimmer, K.G. (1961) Studies on Quantitative Radiation Biology. Radiation Biology Oliver and Boyd, Edinburgh and Long.

Zirkle, R.E. (1957) Partial Cell Irradiation. Advances in Biological and Medical Physics V Acad. Press, NY : 103-146.

Subject Index

δ ray effect 133
μ randomness 105, 218
σ-randomness 218

A 150 135
Absorbed dose 6, 15
Absorbed dose rate 16
Absorption 50
Ambient dose equivalent 336
Amorphous track 226
Amplitude 87
Area monitoring 335
Associated surface 219
Associated volume 217
Atomic orbitals 100
Attenuation coefficient 47
Auger electron 60
Autonomous 269
Autonomous response 265
Avogadro's number 48

Barn 46
Bayes's theorem 245
Bayesian methods 244
Bethe ridge 54
Bhabha 68
Born approximations 87
Bremsstrahlung 68

Cell killing 288
Cema 50
Center-of-mass 52

Change 10
Charged particle equilibrium 49, 213
Chromatin 283
Chromosomes 283
Chromosome breaks 24, 285
Cislesions 302
Classical electron radius 62
Collective (coherent) excitations 69
Collision 73
Compound dual radiation action 303
Compound lesions 293, 304
Compton scattering 58
Conduction states 100
Continuous slowing down 213
approximation
Convolution theorem 33
"counter" dose equivalent 340
Convolutions 31
Critical charge 368
Critical lesions 315
Cross section 45
Crossers 131, 215
CSDA 213
Cumulative distribution function 193

Delocalized 70
Delta ray equilibrium 50
Delta rays 10
Deposition 43
Dielectric response function 82, 92
Direct effect 357
Direct hits 283
Directional dose equivalent 337
Discrete Fourier transforms 241
Distance model 297

Dose average 30
Dose distribution in y 38
Dose equivalent 331
Dose equivalent index 335
Dose modifying factor 291
Dose rate 279
Dose-averaged energy deposited 192
Double differential cross section 88
Double-strand breaks 303
Drude formula 98
Drude-function 98
Dry DNA 284
DSB 303
Dual radiation action 168
Dynamic form factor 92

Effect 10
Eigenvalues 83
Eigenvectors 83
Elastic collisions 53
Elastic scattering 203
Electric breakdown 125
Encounter radius 361
End-points 280
Endothermic 54
Energy absorbed 39
Energy bands 99
Energy deposit 10
Energy imparted 28
Energy loss function 70, 97
Energy transfer 222
Energy transfer points 10
Entropy 246
Equivalent dose 333
Eucaryotic 283
Event 12, 29
Event frequency 19, 30, 269
Event size 327
Event-by-event 201
Excitation 207
Excitons 69
Exothermic 54
Expectation values 15

Fano factor 123
Fast neutrons 64
First-order 276
Fluence 45, 213
Fluorescence yield 60
Force of mortality 267
Form factor 90, 92
Fourier transforms 90
Fluorescence 60
Free path 50
Frequency average 30
Full width at half maximum 149

Galactic cosmic ray 371
GCR 371
Generalized oscillator strength 71
Glancing collisions 44
GOS 71
Gray 28
Green function 94
Green's function 86
Gross sensitive volume 268
GSV 268

Half-value layer 51
Hamiltonian operators 83
Hard collisions 44
Hartree-Fock 69
Hazard function 266
Histones 283
Hit-size effectiveness 347

ICRU sphere 335
Inactivating events 272, 274
Inactivation 280
Inchoate distribution 7
Indirect action 357
Indirect effect 335
Inelastic collisions 53
Inelastic scattering 92
Informational content 250
Insiders 131, 216

Inter-track 295
Intra-track 295
Interaction potential 83
Intermediate neutrons 64
Internal randomness 371
Ionization 205
Ionization potential 9
Ionizations 39
Ionizing radiation 8

Kellerer-Hug theorem 274
Kerma 48
Kerma factor 49
Klein-Nishima 61

Laboratory (L) system 52
Lesion 293
Lesions 263
LET 18
LET distribution 110
Likelihood 245
Lineal energy 18, 28
Linear energy transfer 18, 49
Linear representation 37
Linear stopping power 50
Linker 283
Local 269
Log(y) representation 37
Log-log representation 37

Macroscopic cross section 51
Mass attenuation coefficient 47
Mass energy absorption coefficient 48
Mass energy transfer coefficient 48
Mass stopping power 49
Matrix 264
MAXENT 244, 254
Maximum entropy 244
Maximum entropy principle 254
Maxwell's equations 92
Mean chord length 28
Mean energy expended 119

Mean excitation energy 75
Mean free path 51
Mean inactivation dose 272
Mean-free path (mfp) 51
Measure 199
Microdosimetric events 268
Micrometer domain 281
Minidosimetry 280
Misrepair 285
Molar mass 48
Møller 67
Momentum operator 85
Momentum space 85
Momentum transfer 97
Monte Carlo 192
Mortality 267
Multi-event 248
Muscle tissue 134
Mutations 285

Nanodosimetric 315
Nanometer domains 281
Natural units 82
Non-local 92
Non-stochastic 280
Non-stochastic quantity 14
Nucleosomes 283
Nucleotide 282
Number of events 273
Number of ions 122

One-hit 276
Operational quantity 336
Orthogonalized plane 100
waves (OPW)
Oscillator strength 91

Pair production 58
Partition function 257
Penumbra 161
Personal dose equivalent 337
Photo-nuclear reactions 58
Photoelectric effect 58

Plane wave 83
Plasma frequency 71
Plasmons 69
Point-pair distribution 190
of distances
Poisson 78
Poisson-compound process 78
Polarization 121
Polya distribution 130
Position representation 85, 86
Positron annihilation 63
Posterior probability 245
Prescribed diffusion 359
Principal value 97
Prior probability 245
Probability density distribution 14
Probability density function 193
Probability distribution 14
Projectiles 43
Proton edge 147
Proximity function 19

Quality factor 332

Radiation biophysics 263
Radiation chemistry 358
Radiative stopping power 73
Random-phase approximation 99
(RPA)
Range 80, 215
Range straggling 81
Rayleigh 58
Re-entry effect 132
Reactivity 268
Regional 7
Rejection technique 196
Relative steepness 274
Relative variance 109
Relevant (energy) transfer point 9
Relevant transfer points 11, 40
Resonance 65
Restricted linear energy transfer 19
Risk 336

Scattered wave 87
Scattering 51
Schrödinger equation 94
Scissors operator 101
Self-consistent field (SCF) 69
Sensitive regions 21
SEU 367
Significant energy deposit 11
Significant transfer point 11
Single event upset 367
Single-event 29, 315
Single-particle approximation 99
Single-strand breaks 303
Site 13
Sites 7
Site model 293, 368
Sources 43
Specific energy 16, 28
Specific energy (imparted) 28
Specific quality factor 346, 348
Specified 43
Spread Bragg peak 164
SSB 303
Starters 131, 216
Stationary 212
Steepness 276
Stochastic effects 280
Stochastic quantity 14
Stoppers 131, 216
Stopping power 73
Straggling 106
Structural changes 280
Structural microdosimetry 7
Sublesions 293
Sublethal damage 278
Sum rule 91
Sum rules 99
Survival 288
Surviving fraction 276
Synergism 312
Synergistically 290

Target 45
Theoretical resolution 150

Thermal neutrons 64
Threshold 55
Time of flight 153
Track segment approximation 209
Transformation 287
Translesions 302
Transport 44
Transport equation 211
Transport theory 211
Triplet production 63

V effect 133
Variance technique 168
Variance-covariance 169

W 119
Wall effect 131
Wall-less counters 131
Wave packet 84

yf(y) 37

Springer-Verlag
and the Environment

We at Springer-Verlag firmly believe that an international science publisher has a special obligation to the environment, and our corporate policies consistently reflect this conviction.

We also expect our business partners – paper mills, printers, packaging manufacturers, etc. – to commit themselves to using environmentally friendly materials and production processes.

The paper in this book is made from low- or no-chlorine pulp and is acid free, in conformance with international standards for paper permanency.

Druck: Mercedesdruck, Berlin
Verarbeitung: Buchbinderei Lüderitz & Bauer, Berlin